Lecture Notes in Artificial Intelligence 679

Subseries of Lecture Notes in Computer Science
Edited by J. Siekmann

Lecture Notes in Computer Science
Edited by G. Goos and J. Hartmanis

T0222697

C. Fermüller A. Leitsch T. Tammet N. Zamov

Resolution Methods for the Decision Problem

Springer-Verlag
Berlin Heidelberg New York
London Paris Tokyo
Hong Kong Barcelona
Budapest

Series Editor

Jörg Siekmann
University of Saarland
German Research Center for Artificial Intelligence (DFKI)
Stuhlsatzenhausweg 3, D-66123 Saarbrücken 11, FRG

Authors

C. Fermüller
A. Leitsch
Technical University of Vienna
Resselgasse 3/285, A-1040 Wien, Austria

Tanel Tammet
Institute of Cybernetics, Estonian Academy of Sciences
Tallinn 200108, Estonia

Nail Zamov
Department of Mathematics, University of Kazan
Lenin Street 18, 420008 Kazan, Russia

CR Subject Classification (1991): I.2.3, F.4.2-3

ISBN 3-540-56732-1 Springer-Verlag Berlin Heidelberg New York
ISBN 0-387-56732-1 Springer-Verlag New York Berlin Heidelberg

This work is subject to copyright. All rights are reserved, whether the whole or part of the material is concerned, specifically the rights of translation, reprinting, re-use of illustrations, recitation, broadcasting, reproduction on microfilms or in any other way, and storage in data banks. Duplication of this publication or parts thereof is permitted only under the provisions of the German Copyright Law of September 9, 1965, in its current version, and permission for use must always be obtained from Springer-Verlag. Violations are liable for prosecution under the German Copyright Law.

© Springer-Verlag Berlin Heidelberg 1993
Printed in Germany

Typesetting: Camera ready by author
45/3140-543210 - Printed on acid-free paper

PREFACE

This volume contains work on the decision problem done in Kazan(Russia), Tallinn (Estonia) and Vienna (Austria). The authors met several times to discuss and exchange their results and finally decided to write a monograph together. Besides a unified treatment of previously published results there are many new results first presented in this volume. We thank Grigori Mints for enlighting discussions and valuable hints. We are also grateful to Mrs. Franziska Gusel for typing the manuscript.

February 1993 C. Fermüller, A. Leitsch, T. Tammet, N. Zamov

TABLE OF CONTENTS

Chapter 1

INTRODUCTION

It is a central problem of logic and computer science to determine whether a formula in predicate logic is satisfiable. Its roots can be traced back to Leibnitz's brave vision of a calculus ratiocinator [Leib], which would allow to settle arbitrary problems by purely mechanical computation, once they have been translated into a special formalism. The actual history of the modern understanding of this project starts around 1900 when Hilbert defined the problem to find an algorithm which decides the validity of first order predicate logic formulas [Hil01]. He called this decision problem the "fundamental problem of mathematical logic". At that time a positive solution seemed to be merely a question of wealth of mathematical invention.

Although no precise concept of "algorithm" was available during the first three decades of our century, some decidable classes of predicate logic were found during this time. The algorithms provided for these classes were clearly effective in any plausible intuitive sense of the word. One of the first results was the decidability of the monadic class (i.e. the class of first order formulas containing only one-place predicate symbols) [Löw15]. In the same paper Löwenheim showed that dyadic logic gives a reduction class, i.e. a class of first order formulas effectively "encoding" full predicate logic; a decision algorithm for such a reduction class could thus be transformed into a decision algorithm of full predicate logic.

In 1936 A. Church demonstrated rigorously that a positive solution of the decision problem is impossible [Chu36], which means that no algorithm can decide the validity (or the satisfiability) of arbitrary first order formulas. From this result the undecidability of the reduction classes follows immediately. However the search for decidable classes has been continued (indeed it gained additional practical and epistemological importance).

The decision problem with respect to the prefix structure of prenex normal forms is solved completely (see [DG79] and [Lew79]). There are much less results, however, for classes with full functional structure; one of the first results on functional classes was the decidability of the satisfiability problem for formulas with prefix $\forall$ and arbitrary function symbols in the matrix [Gur73]. Recently functional classes have been investigated in papers of the authors ([Fer90], [Lei90], [Tam90]). But rather than finding new decidable classes only, the purpose of this monograph is to analyse the decision methods themselves; as basic technique we use resolution and its refinements and variants. While the book of Dreben and Goldfarb on decidable classes [DG79] is based on model theoretical techniques, our approach follows the proof theoretical tradition. Before Robinson's famous paper on the resolution principle [Rob65] was published, S. Maslov succeeded to prove the decidability of the $\exists^*\forall^*\exists^*$ - Krom class (the so called Maslov class) by proof theoretical means [Mas64]; he used the inverse method which can be considered as a special version of the resolution method based on sequent calculus. One of the authors used a variant of this method to prove the decidability of Maslov's K-class [Zam72], [Zam89]. A typical feature of this method (and of resolution) is the essential use of most general unifiers; Instead on ground level inference takes place on the predicate logic level with variables (Herbrand's theorem is used only for completeness results - not for decision methods). Further results on decidable classes, based on this method, were obtained by Zamov and Sharonov in 1974 [ZS74].

Independently Joyner showed in his thesis how the resolution principle can be adapted to decide some of the classical prefix classes [Joy73]. His method consists in defining specific A-ordering refinements of resolution, which are complete and produce only finitely many resolvents on certain clause classes (the classes are Skolemized conjunctive normal forms of prefix classes). Although Joyner used A-ordering methods only, his basic idea is quite general: Find a complete resolution refinement which is terminating (i.e. it produces only finitely many resolvents) on a class of sets of clauses Γ. Then this refinement decides Γ, because either a contradiction is derived or the deduction process stops "without success"; in the latter case we have found that the set of clauses is satisfiable (without defining a model). This is the principle on which most of the methods and results of this monograph are based (although for some classes refinements are defined which are complete on the corresponding decision class only).

In chapter 3 (written by A.Leitsch) semantic clash resolution as decision procedure is investigated. Here some earlier results on decidable Horn classes [Fer90], [Lei90] are generalized to clause classes with arbitrary propositional structure. The classes are characterized by term depth and variable-occurrence properties, but there are no restrictions on function symbols (thus the classes cannot be obtained by Skolemization of function-free classes). Instead of fixing a specific semantic refinement, a prover generator is defined, which produces terminating setting refinements out of the syntactical structure of the clause sets. By this technique the classes PVD and OCC1 are shown to be decidable. OCC1 is defined by occurrence restrictions to variables in the (semantically) positive part of the clause; for its decision restricted factoring and condensing are required. PVD is a strong extension of DATALOG and is "sharp" w.r.t. undecidability (a small change in the definition of the class yields the representation of a word problem in an arbitary equational theory). The decision procedure for PVD is shown to give an efficient resolution decision procedure for the Bernays - Schönfinkel class. A subchapter is devoted to a subclass of the Horn clause implication problem; here positive hyperresolution with ordered rule clauses provides a basis for the decision algorithm.

Chapter 4 (written by T.Tammet) gives a detailed treatment of ordering refinements and their completeness properties. Various types of ordering principles (a-priori and a-posteriori orderings) are compared and general completeness conditions for π - orderings are presented. It is shown that all ordering refinements are compatible with backward subsumption (but not with forward subsumption). A- ordering methods appear as specific π -refinements, where the full deletion method is allowed. A short overview (examples) of using ordering refinements as decision algorithms is (are) presented. Finally an ordering, which is not a π -ordering (called v-ordering), is used to decide the class E^+ (containing the one-variable functional clause class). While the v-ordering refinement is shown to terminate on E^+, its completeness (on E^+) is still an open problem. However it is proved that the v-ordering is complete on a subclass of E^+ by a locking technique.

Chapter 5 (written by C. Fermüller) describes various decision procedures based on A-ordering refinements and saturation. To make the chapter more self-contained, there is a detailed presentation of semantic-tree-completeness,

A-orderings and splitting principles. Via the concept of covering terms several A-ordering refinements are shown to terminate on specific clause classes. One of the classes, being decidable this way, is the class E1 which properly contains the extended Ackermann class. For the class E^+ (which is also discussed in chapter 4) an (a-posteriori) ordering is defined which is not an A-ordering. Similar to the v-ordering refinement in chapter 4 this ordering refinement is shown to terminate on E^+. But by combining it with a saturation technique one gets a refinement which also is complete on E^+ (this yields the first full proof of decidability for E^+). Combining A-ordering with saturation gives a decision procedure for the class S^+ which is a generalization of the initially extended Skolem class (and thus contains the Gödel class). A pure saturation method, combined with unrestricted resolution is introduced as decision procedure for a class containing Maslovs $\exists^*\forall^*\exists^*$-Krom class; while the completeness of the method is independent of the propositional clause form, the Krom-property is needed for termination.

In chapter 6 (written by N. Zamov) a decision procedure for Maslov's K-class based on an ordering refinement is presented. The K-class is very general and contains the initially extended Skolem class (but it is not comparable with the S^+-class in chapter 5). A detailed analysis of the behaviour of most general unifiers and ordered resolvents of the class is perfomed by the concept of regular terms and by the domination relation among literals. A specific π-ordering is defined on the ground level which then is lifted to an ordering on the general level. The termination of the resulting ordering refinement on K is proved (the property of regularity is preserved under the ordering refinement) and the completeness of the method is inferred by the completeness results in chapter 4.

In chapter 7 (written by T.Tammet) methods of resolution and narrowing are applied to automatical finite model building. Rather than just testing satisfiability of clause sets by termination of resolution refinements, a method is presented which constructs finite models for the union of the initially extended Ackermann- and the essentially monadic class. First the set of clauses is transformed into a set containing literals only of the essentially monadic type. Then, by use of narrowing on the set of ground terms of the Herbrand universe, a finite domain interpretation of the function symbols is constructed, which then can be extended to an interpretation of the predicate symbols. By its proof theoretical nature, based on tools of computational logic, this method of building finite models is much more efficient than exhaustive search through finite domain interpretations up to some (recursively computable) domain size. A similar method invented by the author [Tam90] was tested on several examples and proved to be very efficient.

In chapter 8 several applications of resolution decision methods are discussed. Consistency is of central importance to terminological logics, such as KL-ONE. It is shown that the language ALC (which can be considered as a basic language for KL-ONE systems) can be translated into the class S^+ (defined in chapter 5); thus A-ordering + saturation gives an algorithm testing consistency of formulas from ALC (there is also an apriori ordering refinement deciding the translation of ALC). It is shown that a variant of ALC allowing arbitrary function symbols and ground equalities can be translated into the one-variable class (which is a subclass of E^+ discussed in the chapters 4 and 5). The following sections are spent on descriptions of experiments with resolution decision procedures as theorem provers. The prover generator, defined for deciding the class PVD in chapter 3 and based on semantic clash resolution, is tested on 3 examples. On the first two examples the PVD-decision method proved to be strongly superior to a semantic clash prover with fixed setting; example 3 could not be handled by any refinement described in the monograph. Also in the case of ordering refinements decision procedures are very efficient theorem provers; so formulas in the classical book of Church [Chu56] could be proved much faster than by other theorem provers (the latter even failed on some examples). The experiments thus indicate that it pays out to incorporate decision methods into general theorem proving. A way to do this, is to design an expert system classifying a set of clauses S before deduction; if it turns out that S is contained in a decidable class Γ then apply the resolution decision procedure for Γ. Although this method cannot always work (by the undecidability of clause logic) it can be quite useful for practical purposes.

Chapter 2

TERMINOLOGY

In this chapter we provide definitions for the basic notions of clause logic and also introduce some more special terminology that we shall use throughout this monograph. Although we assume the reader to be familiar with the concept of resolution we review the fundamental definitions for the sake of clarity and completeness. Additional terminology will be introduced in later chapters whenever this helps the understanding of our definitions and proofs.

2.1 TERMS, LITERALS AND CLAUSES

Concerning the language of clause logic we assume that there is an infinite supply of *variable symbols V*, constant symbols CS, function symbols FS, and predicate symbols PS. As usual we assume each function and predicate symbol to be associated with some fixed arity which we denote by arity(F) for F in PS or FS. We call a predicate or function symbol unary iff it is of arity 1, binary iff the arity is 2, and in general n-place for arity n. The set of n- place function and predicate symbols is denoted by FS_n and PS_n, respectively.

If S is some set of expressions, clauses or clause sets then CS(S), FS(S), and PS(S), refers to the set of constant, function and predicate symbols, respectively, that occur in S. (For a formal definition of occurrence see definition 2.12 below).

We define the notions term, atom, literal, expression and clause formally:

DEFINITION 2.1: A term is defined inductively as follows:

(i) Each variable and each constant is a term.

(ii) If $t_1, \dots, t_n$ are terms and f is an n-place function symbol, then $f(t_1, \dots, t_n)$ is also a term.

(iii) Nothing else is a term.

If a term t is of the form $f(t_1, \dots, t_n)$ we call it functional; the set of arguments of t - args(t) - is $\{t_1, \dots, t_n\}$; f is called the leading (function) symbol of t. The set of all terms is called T.

DEFINITION 2.2.: If $t_1, \dots, t_n$ are terms and P denotes an n-place predicate symbol then $A = P(t_1, \dots, t_n)$ is an atom; P is called the leading (predicate) symbol of A; args(A) is the set $\{t_1, \dots, t_n\}$.

DEFINITION 2.3.: A literal is either an atom or an atom preceded by a negation sign.

DEFINITION 2.4.: An expression is either a term or a literal.

DEFINITION 2.5.: A clause is a finite set of literals. The empty clause is denoted by $\square$.

Throughout this work we shall speak of classes of clause sets, by this we always mean sets of finite sets of clauses.

DEFINITION 2.6.: If a literal L is unsigned, i.e. if it is identical with an atom A, then the dual of L— L^d — equals $\neg A$. Otherwise, if L is of the form $\neg A$ then $L^d = A$. For a set of literals $C = \{L_1, \dots, L_n\}$ we define $C^d = \{L_1^d, \dots, L_n^d\}$.

Additionally we introduce the following notation:

DEFINITION 2.7.: C_+ is the set of positive (unsigned) literals of a clause C, analogously C_- denotes the set of negative literals (negated atoms) in C.

DEFINITION 2.8.: C is a Horn clause iff it contains at most one positive literal, i.e. $|C_+| \leq 1$.
A Horn clause C with $C_+ = C$ is called a fact, with $C_- = C$ a goal; if C_+ and C_- are both nonempty then C is called a rule.

2. 2 TERM STRUCTURE

The term depth of an expression or a clause is defined as follows:

DEFINITION 2.9.: The term depth of a term t - $\tau(t)$ - is defined by:

(i) If t is a variable or a constant, then $\tau(t) = 0$.

(ii) If $t = f(t_1, \ldots, t_n)$, where F is an n-place function symbol, then $\tau(t) = 1 + \max\{\tau(t_i) \mid 1 \le i \le n\}$.

The term depth of a literal L is defined as $\tau(L) = \max\{\tau(t) \mid t \in args(L)\}$. The term depth of a clause C is defined as $\tau(C) = \max\{\tau(L) \mid L \in C\}$. For a set S of clauses we define $\tau(S) = \max\{\tau(C) \mid C \in S\}$.
It is convient to make use of some definitions concerning term structure that are well known form the term rewriting literature.

DEFINITION 2.10.: Let A be an expression. Then the set of positions in A P(A) is a set of sequences of integers, defined as follows:

(i) ε (i.e. the empty sequence) $\in P(A)$;

(ii) if $p \in P(t_i)$ then $i.p \in P(A)$ for $A = F(t_1, \ldots, t_n)$, where F is an n-ary predicate symbol (possibly preceeded by a negation sign) or a function symbol and $1 \le i \le n$;

(iii) nothing else is in P(A).

The length of a position $|p|$ is the number of integers in the sequence. The number of subterms of A - s(A) - is defined to be $|P(A)|$.

DEFINITION 2.11.: For any position p of an expression A the subterm of A at position p SUB(p, A) is defined as follows:

(i) $SUB(\varepsilon, A) = A$;

(ii) $SUB(i.p, A) = SUB(p, t_i)$, if $A = F(t_1, \ldots, t_n)$ for some n-ary predicate symbol (possibly preceeded by a negation sign) or some function symbol F.

The depth of occurence of the subterm of position defined as $\tau_{sub}(p, A) = |p|$.

DEFINITION 2.12.: We say that an expression t occurs in an expression E, iff there is an i, s.t. t = SUB(i, E). Occasionally, we shall write E[t] to indicate that t is a proper subterm of E, i.e. that t occurs in E but $t \neq E$. We also say that a function or predicate symbol F occurs in E iff F is the leading symbol of some expression t that occurs in E.

The set of all variables occurring in E is called V(E); if C is a clause, then V(C) is the union over all $V(P_i)$ for all atoms P_i in C.
We define E_1 and E_2 to be variable disjoint iff $V(E_1) \cap V(E_2) = \emptyset$.
By OCC(x, E) we denote the number of occurrences of a variable x in E, i.e. $OCC(x, E) = | \{ i | SUB(i, E) = x \} |$. OCC(x, C) is defined analogously for clauses C.

DEFINITION 2.13.: An expression or a clause is called ground if no variables occur in it. We call it constant free if no constants occur in it, and function free if it does not contain function symbols.

EXAMPLES: If E = P(x, f(f(y))), then SUB(ε, E) = P(x, f(f(y))), SUB(1.ε, E) = x, SUB(2.ε, E) = f(f(y)), SUB(2.1.ε, E) = f(y), and SUB(2.1.1.ε, E) = y; V(E) = {x, y}. OCC(x, E) = OCC(y, E) = 1. E is not ground, but constant free.

DEFINITION 2.14.: $\tau_{min}(t,E)$ is defined as the minimal depth of occurrence of a term t within an expression E, i.e.

$$\tau_{min}(t, E) = \min\{\tau_{SUB}(i, E)\} \mid SUB(i, E) = t \}.$$

If C is a clause, then $\tau_{min}(t, C)$ denotes the minimum of $\tau_{min}(t, P_i)$ for all atoms P_i of C. $\tau_{max}(t, E)$ respectively $\tau_{max}(t, C)$ are defined in the same way.

EXAMPLES: If P_1 = P(x, f(f(y))), P_2 = Q(f(x)) and C = $\{P_1, \neg P_2\}$, then $\tau(P_1) = 2$, $\tau(P_2 = 1$, $\tau(C) = 2$, $\tau_{SUB}(0, P_1) = \tau_{SUB}(0, P_2) = 0$, $\tau_{SUB}(1, P_1) = 0$, $\tau_{SUB}(3, P_1) = 1$, $\tau_{min}(x, C) = 0$, $\tau_{max}(x, C) = 1$, $\tau_{min}(y, C) = \tau_{max}(y, C) = 2$.

DEFINITION 2.15.: The maximal variable depth of an expression E is defined as $\tau_v(E) = \max\{\tau_{max}(x, E) \mid x \in V(E)\}$. For clauses C we define
$\tau_v(C) = \max\{\tau_v(L) \mid L \in C\}$;
analogously for clause sets S $\tau_v(S) = \max\{\tau_v(C) \mid C \in S\}$.

2.3 SUBSTITUTIONS

Another basic notion is the concept of substitution.

DEFINITION 2.16.: Let V be the set of variables and T be the set of terms. A substitution is a mapping σ: V to T s.t. $\sigma(x) = x$ almost everywhere. We call the set $\{x \mid \sigma(x) \neq x\}$ domain of σ and denote it by dom(σ), $\{\sigma(x) \mid x \in \mathrm{dom}(\sigma)\}$ is called range of σ (rg(σ)). By ε we denote the empty substitution, i.e. $\varepsilon(x) = x$ for all variables x.

We shall occasionally specify a substitution as a (finite) set of expressions of the form t_i/x_i with the intended meaning $\sigma(x_i) = t_i$.

DEFINITION 2.17.: We say that a substitution σ is based on a clause set S iff no other constant and functions symbols besides that in CS(S) and FS(S), respectively, occur in the terms of rg(σ).

A ground substitution is a substitution σ s.t. there are only ground terms in rg(σ).

The application of substitutions to expressions is defined as follows:

DEFINITION 2.18.: Let E be an expression and σ a substitution.

(i) If E is a variable, then $E\sigma$ is $\sigma(E)$ (cf. definition 2.17).

(ii) If E is a constant, then $E\sigma = E$.

(iii) Otherwise E is of the form $F(t_1, \dots, t_n)$, where F is either an p-place function or predicate symbol (possibly negated).

In this case $E\sigma = F(t_1\sigma, \dots, t_n\sigma)$. If L is a literal, then $L\sigma$ is defined to be the application of σ to the atom of L. If C is a set of expressions or a clause, then $C\sigma = \{E\sigma \mid E \in C\}$.

DEFINITION 2.19.: An expression E_1 is an instance of another expression E_2 iff there exists a substitution σ s.t. $E_1 = E_2\sigma$. Likewise a clause C_1 is an instance of clause C_2 iff $C_1 = C_2\sigma$ for some substitution σ.

We may compare expressions, substitutions and clauses using the following ordering relation.

DEFINITION 2.21.: Let E_1 and E_2 be expressions, then $E_1 \leq_s E_2$ - read: E_1 is more general than E_2 - iff there exists a substitution σ s.t. $E_1\sigma = E_2$. For substitutions ρ and ϑ we define analogously: $\rho \leq_s \vartheta$ iff there exists a substitution σ s.t. $\rho\sigma = \vartheta$. Similarly, if C and D are clauses, $C \leq_s D$ iff there exists a substitution σ s.t. $C\sigma \subseteq D$. In this case we also say, in accordance with the usual resolution terminology, that C subsumes D.

DEFINITION 2.22.: A clause is called condensed if it does not subsume a proper subclause of itself.

EXAMPLE:

$\{P(x, a), P(a, x)\}$ is condensed; $C = \{P(x), P(a)\}$ is not condensed, because it subsumes $\{P(a)\}$ which is a subclause of C. For every clause C there is (up to renaming) a unique subclause D s.t. $C \leq_s D$ and D is condensed. We call D the condensation of C and denote it by C_{cond}.

Condensation is an important technique in Joyner's resolution decision procedures [Joy 76].

DEFINITION 2.23.: A set of expressions M is unifiable by a substitution σ iff $E_i\sigma = E_j\sigma$ for all $E_i, E_j \in M$. σ is called most general unifier (m.g.u.) of M iff for every other unifier ρ of M: $\sigma \leq_s \rho$.

We shall also say that E_1 is unifiable with E_2 iff $\{E_1, E_2\}$ is unifiable.

Remember that any two different m.g.u.s of a set of expressions only differ in the names of the variables.

2.4 FACTORS AND RESOLVENTS

DEFINITION 2.24.: A factor of a clause C is a clause $C\vartheta$, where ϑ is a m.g.u. of some $C' \subseteq C$. In case $|C\vartheta| < |C|$ we call the factor non-trivial.

For the resolvent we retain the original definition of Robinson [Rob 65], which combines factorization and (binary) resolution. But be aware that, in some of the chapters to come, we shall locally define clauses and resolvents differently (namely as lists of literals). It will always be clear from the context which concept we are using.

DEFINITION 2.25.: If C and D are variable disjoint clauses and M and N are subsets of C and D respectively, s.t. $N^d \cup M$ is unifiable by the m.g.u. ϑ, then $E = (C - M)\vartheta \cup (D - N)\vartheta$ is a (Robinson)-resolvent of C and D.

If M and N are singleton sets then E is called binary resolvent of C and D.

The atom A of $(N^d \cup M)\vartheta$ is called the resolved atom. We also say that E is generated via A. The elements of N and M are called the literals resolved upon.

DEFINITION 2.26.: For a clause set S we define Res(S) as the set of Robinson-resolvents of S. Additionally we define:

$R^0(S) = S$,

$R^{i+1}(S) = R^i(S) \cup Res(R^i(S))$, and

$R^*(S) = \bigcup_i R^i(S)$.

We say that a clause C is derivable from a clause set S iff $C \in R^*(S)$.

In the following chapters we shall introduce various refinements of Robinson's resolution procedure. By a refinement of resolution we mean an operator Res' s.t. $Res'(S) \subseteq Res(S)$ for all clause sets S. R'^i and R'^* are defined in the obvious way.

In contrast to resolution refinements we shall also define variants of resolution: For resolution variants we allow ordinary resolvents to be replaced by certain instances of it. This technique is also called saturation. (See especially chapter 5 for examples of this method).

2.5 A UNIFICATION ALGORITHM

Some critical parts of proofs in the following chapters demand for a careful tracing of the unification procedure. For this purpose we state a simple version of an unification algorithm. We first have to introduce some additional terminology, which will also prove useful in later sections.

<u>DEFINITION</u> 2.27.: Let E_1, E_2 be two expressions. The set of <u>corresponding pairs</u> $CORR(E_1, E_2)$ - is defined as follows:

(i) $(E_1, E_2) \in CORR(E_1, E_2)$.

(ii) If $(F_1, F_2) \in CORR(E_1, E_2)$ s.t. $F_1 = F(s_1, \ldots, s_n)$ and $F_2 = F(t_1, \ldots, t_n)$, for some function or predicate symbol F (possibly preceded by a negation sign), then $(s_i, t_i) \in CORR(E_1, E_2)$.

(iii) Nothing else is in $CORR(E_1, E_2)$.

A pair $(F_1, F_2) \in CORR(E_1, E_2)$ is called irreducible iff the leading symbols of F_1 and F_2 are different. (F_1, F_2) is called strongly irreducible iff it is irreducible and both, F_1 and F_2, are not variables.

We are now able to present a unification algorithm which finds a m.g.u. ϑ of two variable disjoint expressions E_1 and E_2, if there is one.

```
begin
   ϑ := ε;  i := 0;
   while E₁ϑ ≠ E₂ϑ do
      if CORR(E₁ϑ, E₂ϑ) contains a strongly irreducible pair then failure;
      if there is a pair (s, t) ∈ CORR(E₁ϑ, E₂ϑ) s.t.
         s ≠ t and either s or t is a variable then
         i := i + 1;
         if s is a variable then xᵢ := s; tᵢ := t;
         else {t is a variable}  xᵢ := t; tᵢ := s;
         endif
      else failure
      endif;
      if xᵢ occurs in tᵢ then failure
      else ϑ := ϑ{tᵢ/xᵢ}
      endif
   endwhile
   {ϑ is the m.g.u. of E₁ and E₂ }
end.
```

Remarks: Termination of this algorithm with failure means that there doesn't exist a m.g.u. of E_1 and E_2.

The substitution component t_i/x_i is called the ith mesh substituent of ϑ.

The presented algorithm is by no means well suited for actual computation but it fits nicely purposes of argumentation in some of our proofs.

2.6 SPLITTING

In later chapters, we sometimes refer to the well known splitting rule. To make the argumentation more precise we present a formal definition.

DEFINITION 2.28.: A clause C is called decomposed iff for all subsets C', C" of C s.t. $C' \cup C'' = C$ and both, C' and C", are nonempty:

$$V(C') \cap V(C'') \neq \emptyset.$$

Otherwise, if $V(C') \cap V(C'') = \emptyset$, we say that C can be decomposed into (split components) C' and C".

DEFINITION 2.29.: For any clause set S let SPLIT(S) denote the set of clause sets obtained by splitting all members of S as far as possible. More accurately we define recursively:

(i) $\Sigma_0 = \{S\}$.

(ii) If for all $S' \in \Sigma_i$ and all $C \in S'$ C is decomposed then $\Sigma_{i+1} = \Sigma_i$.

(iii) If there is some $S' \in \Sigma_i$ s.t. for some $C \in S'$ C is can be decomposed into C' and C" then

$$\Sigma_{i+1} = (\Sigma_i - \{S'\}) \cup ((S' - \{C\}) \cup \{C'\}) \cup ((S' - \{C\}) \cup \{C''\}).$$

(A proper C is chosen nondeterministically)

Now we may define $SPLIT(S) = \Sigma_k$ where $k = \min\{m \mid \Sigma_m = \Sigma_{m+1}\}$.

The splitting rule says that we have to apply resolution to all members of SPLIT(S) separately to test for the unsatisfiablity of S. (Recall that S is unsatisfiable iff S' is unsatisfiable for all S' ∈ SPLIT(S).)

2.7 HERBRAND SEMANTICS

For the semantics of clause logic we refer, as usual, to the terminological machinery developed by J. Herbrand. We review the basic definitions:

DEFINITION 2.30.: The Herbrand universe H_S of a clause set S is the set of all ground terms s.t. only constant and function symbols in CS(S) and FS(S) occur in them. (If CS(S) is empty we introduce a special constant symbol to prevent H_S from being empty).

DEFINITION 2.31.: The Herbrand tree HT_S of a clause set S is a directed graph of the following structure: The vertices of HT_S are identified with the elements of H_S; there is an edge from the vertex s to the vertex t ($s,t \in H_S$) iff $t \in args(s)$.

Observe that HT_S does not contain directed cycles.

DEFINITION 2.32.: An Herbrand instance of an atom or a clause C in S is a ground instance $C\vartheta$ of C s.t. ϑ is based on S.

DEFINITION 2.33.: The Herbrand base $\hat{H}_S$ is the set of all Herbrand instances of atoms appearing in clauses of S.

DEFINITION 2.34.: An Herbrand interpretation HI_S for a clause set S is a subset of $\hat{H}_S$ with the intended meaning that the truth value true is assigned to all elements of HI_S and the truth value false is assigned to all atoms in $\hat{H}_S - HI_S$.

Remark: We shall also denote HI_S as

$$\{A \mid A \in HI_S\} \cup \{\neg A \mid A \in \hat{H}_S - HI_S\}.$$

2.8 REMARKS

Of course, clause logic may be viewed just as a special syntactic frame for more "classic" formulations of first order predicate logic. We assume familiarity with such formulations and refer to the standard textbooks for such notions as first order formula, quantificational prefix, conjunctive normal form, prenex normal form etc..

Recall that the correspondence between first order formulas and clause sets is established via transformation of formulas to prenex normal form and Skolemization (i.e. eliminating existential quantifiers by substituting certain functional terms for the corresponding variables).

Algorithms for the transformation of a formula to some clause set which is equivalent w.r.t. satisfiability can be found in textbooks on automated theorem proving [CL 73].

Throughout this work we make use of the following naming conventions: For variable symbols we use letters from the end of the alphabet (u, v, w, x, y, z); for constant symbols, letters a, b, c are used; function symbols are denoted by f, g or h; as metavariables for terms we use t or s; capital letters will denote atoms, literals, clauses or certain sets of expressions. Whenever needed this letters are augmented by indices.

Chapter 3

SEMANTIC CLASH RESOLUTION AS DECISION PROCEDURE

3.1 INTRODUCTION AND DEFINITIONS

Semantic resolution was one of the first resolution refinements developed after Robinson's famous paper on resolution [Rob65a]. Already in the same year, this author proposed the refinement of hyperresolution [Rob65b]; even somewhat earlier S. Maslov [Mas64] used a clash method defined within sequent calculus to decide the class $\exists^*\forall^*\exists^*$Krom. The essential feature of clash methods is the possibility of "macro-inference", that is performing more than one resolution step at once and forgetting the intermediate results. Still hyperresolution is one of the most efficient resolution refinements. In [Sla67] Slagle generalized Robinson's idea to what is generally known as semantic clash resolution (RSC). The basic idea is to choose a Herbrand model $\mathfrak{m}$ and separate true clauses (in $\mathfrak{m}$) from false clauses (in $\mathfrak{m}$); true clauses may only be resolved with false clauses. Even with the clash property and with restricted factoring, the method is complete and can be combined with the deletion strategy (tautology-elimination + subsumption).

In this chapter we show that semantic clash resolution is well suited to decide a wide range of clause classes. That means we can give some general syntactic criteria on clauses which guarantee termination of RSC for an appropriate setting (which can be constructed effectively). Particularly we generalize some results on decidable Horn classes, which were based on hyperresolution and unit resolution ([Fer90], [Lei90]). The resolution refinements used in this chapter do not only decide interesting clause classes, but at the same time yield quite efficient theorem provers (because they are complete). We also show that the decidable classes are quite "sharp" in the sense that slight weakening of restrictions already gives undecidable classes.

We now give some basic definitions.

DEFINITION 3.1.1.: Let $\mathcal{C}$ be a set of clauses and $\{P_1, \dots, P_n\}$ the set of predicate symbols occuring in $\mathcal{C}$. A setting $\mathfrak{m}$ is a Herbrand interpretation which, for every P_i, assigns all ground atoms $P_i(\bar{t})$ to true or all $P_i(\bar{t})$ to false ($\bar{t}$ is a ground term vector of appropriate arity).

Remark: The definition above is not the most general one, but suffices for our purposes; thus we follow the definitions of Chang & Lee ([CL 73]) rather than that of Loveland ([Lov 78]).

EXAMPLE 3.1.1.:

$\mathcal{C} = \{\{P(b)\}, \{\neg P(x), P(f(x))\}, \{\neg P(f(f(a)))\}\}$; $H = \{a, b, f(a), (f(b), \dots\}$

$\mathfrak{m}_1 = \{P(t) \,/\, t \in H\}$, $\mathfrak{m}_2 = \{\neg P(t) \,/\, t \in H\}$ are the only two possible settings. $\mathfrak{m} = \{\neg P(a), P(b), \neg P(f(a)), P(f(b)), \neg P(f^2(a)), \dots\}$ is a Herbrand interpretation, but is not allowed as a setting. With respect to $\mathfrak{m}_1$, $\{P(b)\}$ and $\{\neg P(x), P(f(x))\}$ are true in $\mathfrak{m}_1$ (every ground instance is true in $\mathfrak{m}_1$), but $\{\neg P(f^{(2)}(a))\}$ is false $\mathfrak{m}_1$.

DEFINITION 3.1.2.: Let $\mathcal{C}$ be a set of clauses and $\mathfrak{m}$ be a setting for $\mathcal{C}$. A resolvent of two clauses C_1, C_2 in $\mathcal{C}$ is called an $\mathfrak{m}$ - resolvent if one of C_1, C_2 is false in $\mathfrak{m}$ (note that, by definition of a setting, no resolvent exists if both clauses are false in $\mathfrak{m}$).

For the set of clauses $\mathcal{C}$ in example 3.1.1. there is only one possible $\mathfrak{m}_1$ - resolvent; this one is $\{\neg Pf(a))\}$ (resolve the second with the third clause). After this resolution we get

$$\mathcal{C}_1 = \{ \underset{C_1}{\{P(b)\}}, \underset{C_2}{\{\neg P(x), P(f(x))\}}, \underset{C_3}{\{\neg P(f(a))\}}, \underset{C_4}{\{\neg P(f^2(a))\}}\}$$

While resolvents from (C_2, C_3) and (C_2, C_4) are allowed now, the only new one is $\{\neg P(a)\}$ (derived from (C_2, C_3)) and we get $\mathcal{C}_2 = \{C_1, C_2, C_3, C_4, \{\neg P(a)\}\}$. It is easy to verify that no new resolvents are derivable from $\mathcal{C}_2$. But $\mathcal{C}_2$ does not contain $\square$; because $\mathfrak{m}$ - resolution is complete (a consequence of the completeness of RSC [CL 73]) we have proved that $\mathcal{C}$ is satisfiable. Starting with setting $\mathfrak{m}_2$ we would not do so well anymore; in this case we could derive the sequence $\{P(b)\}$, $\{P(f(b))\}$, $\{P(f^2(b))\}$, ... and resolution would never stop on $\mathcal{C}$. While $\mathfrak{m}_1$- and $\mathfrak{m}_2$ - resolution are both complete, $\mathfrak{m}_1$ - resolution decides $\mathcal{C}$ (that means it is terminating on $\mathcal{C}$), but $\mathfrak{m}_2$- resolution does not. Later we will show that for some classes appropriate settings (which yield termination) can be found automatically by an analysis of the term- and variable structure in clauses.

DEFINITION 3.1.3.: An $\mathfrak{m}$ - resolvent of clauses C_1, C_2 s.t. C_1 is true in $\mathfrak{m}$ and C_2 is false in $\mathfrak{m}$, is said to be defined "under restricted factoring" from C_1, C_2 if factoring substitutions are only allowed in C_2.

EXAMPLE 3.1.2.:

$$\mathcal{C} = \{\underbrace{\{P(x),\ P(y)\}}_{C_1},\ \underbrace{\{\neg P(u),\ \neg P(v)\}}_{C_2}\},\qquad \mathfrak{m} = \{P(a)\}.$$

Let $A = \{P(y)\}$, $B = \{\neg P(u),\ \neg P(v)\}$ and $\lambda = \{y/u,\ y/v\}$.
Then $\{P(x)\} = (C_1 - A)\lambda \cup (C_2 - B)\lambda$ is defined under restricted factoring.
If, on the other hand, we set $A = \{P(x),\ P(y)\}$, $B = \{\neg P(u)\}$ and $\lambda = \{u/x, u/y\}$ then $(C_1 - A)\lambda \cup (C_2 - B)\lambda$ $(= \{\neg P(v)\})$ is not defined under restricted factoring.

DEFINITION 3.1.4.: A semantic clash sequence is a sequence Γ of the form $(C;\ D_1, \dots, D_n)$ where $C, D_1, \dots, D_n$ are clauses in some set of clauses $\mathcal{C}$, C is true and all D_i are false in a setting $\mathfrak{m}$. C is called nucleus, the D_i electrons of Γ.
Let $R_0 = C$ and

R_{i+1} = an $\mathfrak{m}$ - resolvent of R_i and D_{i+1} defined under restricted factoring (if it exists).

If R_n (defined after a linear deduction of n resolution steps) is false in $\mathfrak{m}$ (it might be $\square$) then R_n is called semantic clash resolvent of Γ with respect to $\mathfrak{m}$ (SC$\mathfrak{m}$ - resolvent).

EXAMPLE 3.1.3.:

Let
$$\mathcal{C} = \{\underbrace{\{P(x),\ Q(x,y),\ \neg Q(x,g(y)),\ \neg Q(x,f(y))\}}_{C_1},\ \underbrace{\{Q(u,v),\ R(v)\}}_{C_2},\ \underbrace{\{Q(u',g(u')),\ R(g(u'))\}}_{C_3}\}$$

$\Gamma = (C_1;\ C_2,\ C_3)$ is a clash sequence w.r.t. $\mathfrak{m} = \{\neg P(s),\ \neg Q(s,t),\ \neg R(s) \mid s,t \in H(\mathcal{C})\}$.

$R_0 = C_1$

$R_1 = \{P(x), Q(x,y), R(f(y)), \neg Q(x,g(y))\}$

$R_2 = \{P(x), Q(x,y), R(f(y)), R(g(y))\}$

R_2 is false in $\mathfrak{m}$ and thus is a SC$\mathfrak{m}$ - resolvent of Γ.

Note that there are two possibilities to define a resolvent R_1; but by resolving the literals $\neg Q(x,g(y))$, $Q(u,v)$ we get $R_1' = \{P(x), Q(x,y), R(g(y)), \neg Q(x,f(y))\}$.

But R_1 cannot be resolved with C_3. We get that Γ defines only one clash resolvent in this case; in general there are several possibilities to define the R_i and there are several resolvents of a clash.

While the completeness of semantic clash resolution with literal ordering in the electrons is contained in the standard literature on automatic theorem proving [CL 73], [Lov78], the completeness of RSC$\mathfrak{m}$ with restricted factoring (as in definition 3.1.3.) was shown by H. Noll in 1980 [Nol 80]. In this chapter we need a further refinement to decide some clause classes; in order to control the growth of the clash resolvents we have to keep all resolvents in condensed form. For definition and properties of condensing see chapter 2.

DEFINITION 3.1.5.: A semantic clash $\Gamma = (C; D_1, \ldots, D_n)$ is called condensed if C and all clauses D_i are condensed.

Like in other papers on resolution decision procedures we don't look at the deductions themselves, but rather at the set of clauses derivable (by some refinement) from the set $\mathcal{C}$.

DEFINITION 3.1.6.: Let $\mathcal{C}$ be a set of clauses and $\mathfrak{m}$ be a setting for $\mathcal{C}$. We define the operators RSC$\mathfrak{m}$D, $R^i_{\mathfrak{m}}$, $R^*_{\mathfrak{m}}$ as follows: RSC$\mathfrak{m}$D($\mathcal{C}$) is the set of all clash resolvents definable by clashes from $\mathcal{C}$ which are represented in condensed form.

More formally:

RSC$\mathfrak{m}$D($\mathcal{C}$) = the set of all C_{cond} s.t. C is a SC$\mathfrak{m}$ - resolvent of a semantic clash in $\mathcal{C}$ according to definition 3.1.4. under equivalence relation $\sim_v$, defined as $C \sim_v D$ iff C is a variant of D; clearly $\sim_v$ is an equivalence relation on the set of clauses.

Furthermore we define recursively:

$\mathcal{C}_0 = \mathcal{C}_{cond} = \{ C \,/\, C \in \mathcal{C},\ C \text{ true in } \mathfrak{m} \} \cup \{ C_{cond} \,/\, C \in \mathcal{C},\ C \text{ false in } \mathfrak{m} \}$.

$\mathcal{C}_{i+1} = (\mathcal{C}_i \cup RSC\mathfrak{m}D(\mathcal{C}_i))/_{\sim_v}$ and $R^i_{\mathfrak{m}}(\mathcal{C}) = \mathcal{C}_i$, $\quad R^*_{\mathfrak{m}}(\mathcal{C}) = \bigcup_{i \in N} R^i_{\mathfrak{m}}(\mathcal{C})$.

By definition, $R^*_{\mathfrak{m}}(\mathcal{C})$ is the set of all clauses derivable by RSC$\mathfrak{m}$D from $\mathcal{C}$. Correctness of SC$\mathfrak{m}$D - resolution means that $R^*_{\mathfrak{m}}(\mathcal{C})$ contains $\square$ only if $\mathcal{C}$ is unsatisfiable; completeness means that the unsatisfiability of $\mathcal{C}$ implies $\square \in R^*_{\mathfrak{m}}(\mathcal{C})$ (for every set of clauses $\mathcal{C}$ and every setting $\mathfrak{m}$ for $\mathcal{C}$).
We now prove the completeness of RSC$\mathfrak{m}$D on the basis of the completeness of RSC$\mathfrak{m}$ (which follows from [Nol80]).

THEOREM 3.1.1.: SC$\mathfrak{m}$D - resolution is complete. More formally:
Let $\mathcal{C}$ be an unsatisfiable set of clauses and $\mathfrak{m}$ be a setting for $\mathcal{C}$; then $\square \in R^*_{\mathfrak{m}}(\mathcal{C})$.

Proof:
Let $S^*_{\mathfrak{m}}(\mathcal{C})$ be the set of clauses deducible from $\mathcal{C}$ by RSC$\mathfrak{m}$ without condensing ($S^*_{\mathfrak{m}}$ can be defined formally like in definition 3.1.6 omitting the condensation operator).
We show that for every $C \in S^*_{\mathfrak{m}}(\mathcal{C})$ there is a $C' \in R^*_{\mathfrak{m}}(\mathcal{C})$ s.t. $C' \leq_s C$.
Because $\square$ is subsumed only by $\square$ itself, $\square \in S^*_{\mathfrak{m}}(\mathcal{C})$ immediately implies $\square \in R^*_{\mathfrak{m}}(\mathcal{C})$ and the completeness of RSC$\mathfrak{m}$D follows from that of RSC$\mathfrak{m}$.
We proceed by induction on i for $S^i_{\mathfrak{m}}(\mathcal{C})$.
$i = 0$: $\quad S^i_{\mathfrak{m}}(\mathcal{C}) = \mathcal{C}$. $R^i_{\mathfrak{m}}(\mathcal{C}) = \mathcal{C}_{cond}$.

By definition of the condensation operator we have $C_{cond} \leq_s C$ for all C and thus for every $C \in S^i_{\mathfrak{m}}(\mathcal{C})$ there is a $C' \in R^i_{\mathfrak{m}}(\mathcal{C})$ with $C' \leq_s C$ for $i = 0$.

(IH) Suppose that for every $C \in S^i_{\mathfrak{m}}(\mathcal{C})$ there is a $C' \in R^i_{\mathfrak{m}}(\mathcal{C})$ s.t. $C' \leq_s C$.

Case i + 1:

Let $C \in S^{i+1}_{\mathfrak{m}}(\mathcal{C}) - S^i_{\mathfrak{m}}(\mathcal{C})$. Then C is resolvent of a clash $\Gamma = (E; D_1,\dots,D_n)$. That means $C = R_n$ for R_j defined as in definition 3.1.4. By (IH) there are clauses D_j' for $j = 1,\dots,n$ s.t. $D_j' \in R^i_{\mathfrak{m}}(\mathcal{C})$ and $D_j' \leq_s D_j$ and E' s.t. $E' \leq_s E$. We show now that either $D_j' \leq_s C$ for some $j \in \{1,\dots,n\}$ or there is a clash resolvent C' out of $\{E', D_1', .., D_n'\}$ s.t. $C' \leq_s C$; the corresponding clash Γ' may be shorter than Γ.

Again, we use induction on j.

$R_0(\Gamma) = E$, $R_0' = E'$ and $E' \leq_s E$ by (IH).

(IH*) Suppose that either $R_k' \leq_s R_j(\Gamma)$ or $D_k \leq_s C$ for some $k \leq j$.

The case for $j = 0$ is obviously valid.

Case j + 1 :

a) $D_k \leq_s C$ for some $k \leq j$.

b) $R_k' \leq_s R_j(\Gamma)$.

In case a) (IH*) is trivial for j+1, because then $D_k \leq_s C$ for some $k \leq j+1$.

b) For simplicity we write $R_j = R_j(\Gamma)$.

Let $R_{j+1} = (R_j - \{L\})\sigma \cup (D_{j+1} - A)\sigma$. s.t. σ is m.g.u. of $\{L^d\} \cup A$ (note that factoring may only occur in D_{j+1} and thus $\{L\}$ must be a singleton set).

By (IH) we have $D_{j+1}' \leq_s D_{j+1}$ and by (IH*) $R_k' \leq_s R_j$.

Suppose that either $R_k' \leq_s R_j - \{L\}$ or $D_{j+1}' \leq_s D_{j+1} - A$;

If $R_k' \leq_s R_j - \{L\}$ then, by $R_j - \{L\} \leq_s R_{j+1}$ we get $R_k' \leq R_{j+1}$.

If $D_{j+1}' \leq_s D_{j+1} - A$ then $D_{j+1}' \leq_s R_l$ for every $l \geq j+1$.

This holds, because all literals in $(D_{j+1} - A)\sigma$ are false in $\mathfrak{m}$ and thus cannot be resolved away by further resolutions in the clash. Thus, in R_n there must be a subclause of the form $(D_{j+1} - A)\sigma\vartheta$, where ϑ is a substitution composed from the m.g.u.'s of the further resolvents in the clash.

It remains to settle the case where $R_k' \not\leq_s R_j - \{L\}$ and $D_{j+1}' \not\leq_s D_{j+1} - A$. Then there is a substitution η s.t. $R_k'\eta \subseteq R_j$ and $B\eta = \{L\}$ for a set $B \subseteq R_k'$. Moreover, there is a substitution ϑ and a set $F \subseteq D_{j+1}'$ s.t. $F\vartheta \subseteq A$, $D_{j+1}'\vartheta \subseteq D_{j+1}$. Furthermore suppose that B and F are maximal, that means $(R_k' - B)\vartheta \cap \{L\} = \emptyset$. Then (supposing that D_{j+1}' and R_k' do not share variables) $B \cup F^d$ is unifiable by a m.g.u. σ'; by $F \leq_s A$, $B \leq_s \{L\}$ it must hold $\sigma' \leq_s \sigma$. Thus we define $R_{k+1}' = (R_k' - B)\sigma' \cup (D_{j+1}' - F)\sigma'$; clearly R_{k+1}' is an $\mathfrak{m}$ - resolvent of R_k' and D_{j+1}', but not (in general) under restricted factoring.

$$\text{Thus } R_{j+1} = (R_j - \{L\})\sigma \cup (D_{j+1} - A)\sigma = ((R_j - \{L\}) \cup (D_{j+1} - A))\sigma =$$
$$((R_k' - B)\eta \cup (D_{j+1}' - F)\vartheta)\sigma = ((R_k' - B) \cup (D_{j+1}' - F))\mu\sigma$$

for some substitution μ.

But $R_{k+1}' = ((R_k' - B) \cup (D_{j+1}' - F))\sigma'$ where σ' is a m.g.u. of $B \cup F^d$. Thus we get $R_{k+1}' \leq_s R_{j+1}$.

Concluding the induction proof (•) we get that there is a clash resolvent C' out of $\{E', D_1', .., D_n'\}$ without restriction of factoring s.t. $C' \leq_s C$ or $D_j' \leq_s C$ for some j.
By lemma 3.1.1. (to be proved afterwards) the clash resolvent C' can be obtained from a clash out of $\{E', D_1', \ldots, D_n'\}$ under restricted factoring (the clash under restriction of factoring may then be longer than the original one). More exactly, we derive from lemma 3.1.1. that there is a

$C'' \in RSC\mathfrak{m}(\{E', D_1', \ldots, D_n'\})$ s.t. C' is a factor of C".

But then $C'' \leq_s C' \leq_s C$ and C''_{cond}, which is in $RSC\mathfrak{m}D(\{E', D_1', \ldots, D_n'\})$, subsumes C by $C'' =_s C''_{cond}$.
We conclude that $C''_{cond} \in R^{i+1}_{\mathfrak{m}}(\mathcal{C})$ and $C''_{cond} \leq_s C$.
In any case we have settled "i+1" of the main induction proof and thus for every $E \in S^{i+1}_{\mathfrak{m}}(\mathcal{C})$ there is a $E' \in R^{i+1}_{\mathfrak{m}}(\mathcal{C})$ s.t. $E' \leq_s E$.

We conclude that for every $E \in S^{*}_{\mathfrak{m}}(\mathcal{C})$ there is an $E' \in R^{*}_{\mathfrak{m}}(\mathcal{C})$ s.t. $E' \leq_s E$.

Q.E.D.

It is easily verified that the proof of theorem 3.1.1. can be modified to a proof of completeness for $SC\mathfrak{m}D$ - resolution with forward subsumption.
Let $RSC\mathfrak{m}$ the same operator as $RSC\mathfrak{m}D$, but without condensing the resolvents. The following technical lemma was necessary to prove theorem 3.1.1.

<u>LEMMA</u> 3.1.1.: Let $\mathcal{C}$ be a set of clauses and $\mathfrak{m}$ be a setting for $\mathcal{C}$. Then for every $C \in RSC\mathfrak{m}(\mathcal{C})$ without restriction of factoring there is a $C' \in RSC\mathfrak{m}(\mathcal{C})$ s.t. $C' \leq_s C$.

<u>Proof:</u>

By the basic lemma in [Nol 80], for every resolvent C of clauses D, E there is a resolution deduction out of {D, E} with "half-factoring" giving a clause C' s.t. $C'\eta = C$ for some instantiation η. If D is true and E is false in $\mathfrak{m}$, then by restricting factoring to negative clauses, we get in fact a semantic clash resolvent C' under restricted factoring. In theorem 2 [Nol 80] it is shown that this property can be extended to whole clashes in semantic resolution. Thus for every $C \in RSC_{\mathcal{M}}(\mathcal{C})$ without restriction of factoring there is a $C' \in RSC_{\mathcal{M}}(\mathcal{C})$ (with restriction of factoring) s.t. $C'\eta = C$ for some substitution η. Particularly, we have $C' \leq_s C$.

Q.E.D.

Parts of the proof of theorem 3.1.1. can be found in the proofs of subsumption results in [Lov 78]. However, Loveland uses another concept of resolution than we do here. Because there might occur some dangerous side effects due to factoring (see [Lei 89]), we have represented these technical matters in more detail. We also note, that H. Noll [Nol 80] uses Robinson's concept on which also this work is based.

We conclude the chapter with some useful notations and definitions.

<u>DEFINITION</u> 3.1.7.: Let $\mathcal{C}$ be a set of clauses and $\{P_1, \dots, P_n\}$ the set of predicate symbols in $\mathcal{C}$. Let $\mathcal{M}_p = \{P_1(\bar{s}_1), \dots, P_n(\bar{s}_n) / \bar{s}_i$ are $H(\mathcal{C})$ - tupels of appropriate arity$\}$; then $\mathcal{M}_p$ is called the <u>positive setting</u> for $\mathcal{C}$. Similarly $\mathcal{M}_n = \{\neg P_1(\bar{s}_1), \dots, \neg P_n(\bar{s}_n) / \bar{s}_i$ are $H(\mathcal{C})$ - tupels of appropriate arity$\}$ is called the <u>negative setting</u> for $\mathcal{C}$.

Let $\mathcal{M}$ be a setting for $\mathcal{C}$ and $C \in \mathcal{C}$. Then C_{neg} is the maximal subset of C which is false in $\mathcal{M}$, $C_{pos} = C - C_{neg}$. For clauses C which are false in $\mathcal{M}$ we clearly have $C = C_{neg}$.

If $\mathcal{C}$ is a clause set then, under negative setting $C_+ = C_{neg}$, under positive setting $C_- = C_{neg}$. Thus C_+, C_- denote the syntactical positivity and negativity, while C_{neg}, C_{pos} describes the semantical status w.r.t. a setting.

3.2 GENERAL CLAUSE CLASSES

In [Lei 90], [Fer90] hyperresolution has been used as decision procedure for some Horn classes. In this chapter we will present the results obtained in these papers and show that, for many of the classes, the Horn restriction is not essential. Moreover, instead of using different forms of hyperresolution (positive and negative) we define a single resolution refinement, which - under appropriate choice of the setting (to be done algorithmically) - decides all classes in [Lei 90], [Fer 90].

We begin with the Horn classes KI, KII, defined in [Lei 90]. For definitions of Horn clauses, rule, fact, goal see chapter 2.

KI: A set of clauses $\mathcal{C}$ belongs to KI if it holds:

a) $\mathcal{C}$ is Horn

b) If $C = \{P, \neg P_1, \dots, \neg P_n\}$ is a rule clause in $\mathcal{C}$ then $\tau(P) = 0$ and $V(P) \subseteq V(\{\neg P_1, \dots, \neg P_n\})$

c) Facts are ground.

It is easy to see that DATALOG [CGT 90] is a subclass of KI (for DATALOG we also have $\tau(C) = 0$ for all $C \in \mathcal{C}$).
It was proved in [Lei90] that positive hyperresolution always terminates on all $\mathcal{C} \in$ KI and thus gives a decision procedure for KI. Because the fixed ordering of the rule clauses was not used in the proof, it is immediately verifiable that RSC$\mathfrak{m}$D decides KI for negative setting $\mathfrak{m}$ (all ground atoms are set to false).

The second class was KII:

A set of clauses $\mathcal{C}$ belongs to KII if it holds:

a) $\mathcal{C}$ is Horn

b) If $C = \{P, \neg P_1, \dots, \neg P_n\}$ is a rule clause then $\tau(\{\neg P_1, \dots, \neg P_n\}) = 0$ and $V(\{\neg P_1, \dots, \neg P_n\}) \subseteq V(\{P\})$

c) Goals are ground.

It is quite obvious that, in some sense, KII is a mirror image of KI. This feature also applies to positive hyperresolution: While for KI the depth of clash resolvents was monotonically decreasing, it is increasing for KII. Thus RSC$\mathfrak{m}$D does not terminate on KII for negative setting $\mathfrak{m}$; however it was shown, that production of resolvents deeper than $\tau(\mathcal{C})$ is useless and that positive hyperresolution can be turned into a decision algorithm by cutting the term depth at $\tau(\mathcal{C})$. On the other hand , RSC$\mathfrak{m}$D for positive setting $\mathfrak{m}$ always terminates an KII on thus yields a decision procedure. Both KI, KII share some common syntax property which is formulated in the following class:

PVD (positive - variable dominated).

DEFINITION 3.2.1: A set of clauses $\mathcal{C}$ belongs to PVD if it holds:
There exists a setting $\mathfrak{m}$ for $\mathcal{C}$ s.t.

PVD 1) All clauses in $\mathcal{C}$ which are false in $\mathfrak{m}$ are ground.

PVD 2) If C is in $\mathcal{C}$ and C is true in $\mathfrak{m}$ then for $x \in V(C_{neg})$ it holds

$$\tau_{max}(x, C_{neg}) \leq \tau_{max}(x, C_{pos}) \text{ and } V(C_{neg}) \subseteq V(C_{pos}).$$

Note that there is no restriction on clauses containing only literals true in $\mathfrak{m}$. It is easy to see that KI $\cup$ KII $\subseteq$ PVD. For KI we have to choose negative, for KII positive setting. Even if restricted to Horn clauses, KI $\cup$ KII is a proper subset of PVD ($\tau(C_{neg}) = 0$ implies $\tau_{max}(x, C_{neg}) \leq \tau_{max}(x, C_{pos})$ for $x \in C_{neg}$, but -in PVD- $\tau(C_{neg})$ need not be zero).

The term "there exists a setting $\mathfrak{m}$" in definition 3.2.1. could suggest, that although there might be such an $\mathfrak{m}$ we cannot find it. Fortunately this is not the case because, by our definition of setting, there are only 2^n settings, n being the number of different predicate symbols in a set of clauses $\mathcal{C}$. Moreover, after having decided for a particular $\mathfrak{m}$, PVD 1) and PVD 2) can be decided algorithmically. We do not deal with algorithms deciding "$\mathcal{C} \in$ PVD" here, but finding efficient ones is certainly an interesting task.

EXAMPLE 3.2.1.:

$$\mathcal{C} = \Big\{ \underbrace{\{P(x), Q(g(x,x))\}}_{C_1}, \underbrace{\{\neg Q(y), R(x,y)\}}_{C_2}, \underbrace{\{\neg R(a,a), \neg R(f(b),a)\}}_{C_3}, \underbrace{\{R(f(x),y)\}}_{C_4}, \underbrace{\{\neg P(x), \neg P(f(x))\}}_{C_5} \Big\}$$

$\mathcal{C}$ is not Horn and cannot be transformed into Horn by changing the signs of the literals.

The following setting $\mathfrak{m} = \{Q(s), R(s,t), \neg P(s) \mid s,t \in H(\mathcal{C})\}$ "brings" $\mathcal{C}$ into PVD. Indeed,

$(C_1)_{neg} = \{P(x)\}$, $\quad (C_1)_{pos} = \{Q(g(x,x))\}$.

$(C_2)_{neg} = \{\neg Q(y)\}$, $(C_2)_{pos} = \{R(x,y)\}$

$C_3 = (C_3)_{neg} = \{\neg R(a,a), \neg R(f(b),a)\}$

$C_4 = (C_4)_{pos} = \{R(f(x),y)\}$

$C_5 = (C_5)_{pos} = \{\neg P(x), \neg P(f(x))\}$

and PVD 1), PVD 2) are fulfilled for all clauses.

<u>THEOREM</u> 3.2.1.: SC$\mathfrak{m}$D - resolution is a decision procedure for PVD. More precisely:

There is an algorithm defining for every $\mathcal{C} \in$ PVD a setting $\mathfrak{m}$ s.t. $R^{*}_{\mathfrak{m}}(\mathcal{C})$ is finite.

<u>Proof:</u>

Let $\mathcal{C} \in$ PVD; because there are only finitely many settings for $\mathfrak{m}$ and PVD 1), PVD 2) can be decided for every $\mathfrak{m}$, we eventually find the setting $\mathfrak{m}$ fulfilling PVD 1), PVD 2).

We show now that $R^{*}_{\mathfrak{m}}(\mathcal{C})$ is finite.

For this purpose it is enough to show both a) and b):

a) $R^{*}_{\mathfrak{m}}(\mathcal{C}) - \mathcal{C}$ contains ground clauses only

b) For all $E \in R^{*}_{\mathfrak{m}}(\mathcal{C}) - \mathcal{C}$ it holds $\tau(E) \leq d$, where

$d = \max\{\tau(C_{neg}) \mathrel{/} C \in \mathcal{C}\}$.

Because there are only finitely many ground clauses of term depth $\leq d$ over the Herbrand universe of $\mathcal{C}$, a), b) guarantee that $R^{*}_{\mathfrak{m}}(\mathcal{C})$ is indeed finite.

We prove a) and b) for $R^{i}_{\mathfrak{m}}(\mathcal{C})$ using induction on i.

$i = 0$: $R^{0}_{\mathfrak{m}}(\mathcal{C}) - \mathcal{C} = \emptyset$ and a) is trivially true. The same holds for b).

(IH) Suppose that a - i) $R^{i}_{\mathfrak{m}}(\mathcal{C}) - \mathcal{C}$ contains only ground clauses; and

b - i) For all $E \in R^{i}_{\mathfrak{m}}(\mathcal{C}) - \mathcal{C}$ we have $\tau(E) \leq d$;

holds.

a-i+1), b-i+1):

Let $\Gamma = (C; D_1, \dots, D_n)$ be a semantic $\mathfrak{m}$ - clash defined by clauses in $R^{i}_{\mathfrak{m}}(\mathcal{C})$. Because all false clauses in $\mathcal{C}$ are ground by PVD 1) and because, by (IH) all derived (false) clauses are ground, the D_i must be ground.

Let $R_0 = C$

R_{i+1} = an $\mathfrak{m}$ resolvent of R_i and D_{i+1} according to definition 3.1.4.

If Γ is resolvable then there is a clash resolvent R_n.

For every pair (R_i, D_{i+1}) and $0 \leq i < n$ there is a literal $L_i \in D_{i+1}$ and a literal $M_i \in R_i$ s.t. the m.g.u. λ_i of the resolution unifies $\{M_i, L_i^d\}$. Note that no factoring substitution is required in the resolution because D_{i+1} is ground. Moreover all the λ_i are ground substitutions fulfilling $M_i\lambda_i = L_i^d$.

Let $\lambda = \lambda_1 \dots \lambda_n$; because all λ_i are ground, we have $M_i \lambda = M_i \lambda_i = L_i^d$.
Because R_n must be false in $\mathfrak{m}$, $C_{pos} = \{M_0, \dots, M_{n-1}\}$ and

$$C_{pos}\lambda \subseteq \{L_0^d, \dots, L_{n-1}^d\}, \quad R_n = C_{neg}\lambda \cup \bigcup_{i=0}^{n-1} (D_{i+1} - \{L_i\}).$$

By PVD 2) we have $V(C_{neg}) \subseteq V(C_{pos})$ (note that C, as every true clause in $R^*_{\mathfrak{m}}$, must be in $\mathcal{C}$).
Because $C_{pos}\lambda$ is ground, $C_{neg}\lambda$ must be ground too.
Because also $\bigcup^{n-1} (D_{i+1} - \{L_i\})$ is ground, also R_n is ground and a- i + 1) is shown.

By PVD 2) we also have $\tau_{max}(x, C_{neg}) \leq \tau_{max}(x, C_{pos})$ for all $x \in V(C_{neg})$.
Either $\tau(C_{neg}\lambda) = \tau(C_{neg})$ or there is a $x \in V(C_{neg})$
s.t. $\tau(C_{neg}\lambda) = \tau_{max}(x, K) + \tau(x\lambda)$ for some $K \in C_{neg}$.
If $\tau(C_{neg}\lambda) = \tau(C_{neg})$, then by definition of d and by (IH)-b-i) we have $\tau(R_n) \leq d$.

If $\tau(C_{neg}\lambda) = \tau_{max}(x, K) + \tau(x\lambda)$ then we know that
$\tau_{max}(x, C_{neg}) \leq \tau_{max}(x, C_{pos})$ and thus
$\tau_{max}(x, K) + \tau(x\lambda) \leq \tau_{max}(x, C_{pos}) + \tau(x\lambda) \leq \tau(C_{pos}\lambda)$.

It follows $\tau(C_{neg}\lambda) \leq \tau(C_{pos}\lambda)$.

But
$C_{pos}\lambda \subseteq \{L_0^d, \dots, L_{n-1}^d\}$ for $L_i \in D_{i+1}$ and thus
$\tau(C_{pos}\lambda) \leq \max\{\tau(D_i) \,/\, i = 1, \dots, n\}$.
Because, by (IH)-b-i) $\tau(D_i) \leq d$ for all $i = 1, \dots, n$ we also get
$\tau(C_{pos}\lambda) \leq d$ and therefore $\tau(C_{neg}\lambda) \leq d$.

$$\tau(R_n) \leq \max\{\tau(C_{neg}\lambda), \ \tau(\bigcup_{i=0}^{n-1}(D_{i+1} - \{L_i\}))\} \leq d$$

is an immediate consequence.

This concludes the proof of b-i+1).

By the induction rule we get for all $E \in R^*_{\mathfrak{m}}(\mathcal{C})$:

a) E is ground

b) $\tau(E) \leq d$.

Q.E.D.

EXAMPLE 3.2.2.:

$\mathcal{C} = \{\{P(x), Q(g(x,x))\}, \{\neg Q(y), R(x,y)\}, \{\neg R(a,a), \neg R(f(b),a)\}, \{R(f(x),y)\},$
$\{P(x), \neg P(f(x))\}, \{\neg P(a)\}\}.$

We apply RSC$\mathfrak{m}$D to decide $\mathcal{C}$.

First we realize that $\mathcal{C}$ is indeed in PVD; we only have to find the setting $\mathfrak{m} = \{Q(s), R(s,t), \neg P(s) \;/\; s,t \in H(\mathcal{C})\}$.

The next step is the computation of $R^{*}_{\mathfrak{m}}(\mathcal{C})$.

$R^{0}_{\mathfrak{m}}(\mathcal{C}) = \mathcal{C}.$

$R^{1}_{\mathfrak{m}}(\mathcal{C}) = \mathcal{C} \cup \{\{\neg R(a,a)\}, \{\neg Q(a), \neg R(f(b),a)\}, \{\neg Q(a), \neg R(a,a)\}\}$

$R^{2}_{\mathfrak{m}}(\mathcal{C}) = R^{1}_{\mathfrak{m}}(\mathcal{C}) \cup \{\{\neg Q(a)\}\}$

$R^{3}_{\mathfrak{m}}(\mathcal{C}) = R^{2}_{\mathfrak{m}}(\mathcal{C})$ and thus $R^{*}_{\mathfrak{m}}(\mathcal{C}) = R^{2}_{\mathfrak{m}}(\mathcal{C})$.

Because $\square \notin R^{*}_{\mathfrak{m}}(\mathcal{C})$ we know that $\mathcal{C}$ is satisfiable.

Note that choosing an arbitrary setting on $\mathcal{C}$ without syntax analysis of clauses in $\mathcal{C}$ can lead to non-termination. If we take for example $\mathfrak{m}\mathrm{p}$ we can derive $\{\neg P(f^{(i)}(a))\}$ for all $i \geq 0$ and thus $R_{\mathfrak{m}_p}{}^{*}(\mathcal{C})$ is infinite.

By computing $R^{*}_{\mathfrak{m}}(\mathcal{C})$ we do not only know that $\mathcal{C}$ is satisfiable, we are also able to construct a Herbrand model for $\mathcal{C}$.

$R^{*}_{\mathfrak{m}}(\mathcal{C}) = \mathcal{C} \cup \{\{\neg R(a,a)\}, \{\neg Q(a)\}, \{\neg Q(a), \neg R(f(b),a)\}, \{\neg Q(a), \neg R(a,a)\}\}.$

If we perform subsumption in the set of the negative clauses in $R^{*}_{\mathfrak{m}}(\mathcal{C})$ we get $\{\neg R(a,a)\}, \{\neg Q(a)\}$. Thus we know that $R(a,a)$, $Q(a)$ must be set to false. A slight modification of $\mathfrak{m}$ then gives a model $\mathfrak{m}_0$:

$\mathfrak{m}_0 = \{\neg R(a,a), \neg Q(a)\} \cup \{Q(s) \;/\; s \in H - \{a\}\} \cup \{R(s,t) \;/\; s \neq a \text{ or } t \neq a\} \cup$
$\{\neg P(s) \;/\; s \in H\}.$

Apparently, we did not use condensing and restricted factoring in deciding PVD. Instead we know from theorem 3.2.1. that both operations cannot be applied in a nontrivial manner on clauses in $\mathcal{C}$ (note that we did not allow condensing of true clauses). However, we will use both condensing and restricted factoring to decide the second class in this chapter. In order to obtain a decision method for both classes (simultaneously), it is reasonable to choose RSC$\mathfrak{m}$D. In [Fer90] two classes D_2, D_3 were defined and investigated which are structurally similar to KI, KII and also show a similar behaviour w.r.t. decision procedures.

Class D_2:

A set of clauses $\mathcal{C}$ belongs to D_2 if

D_2- a) $\mathcal{C}$ is Horn.

D_2-b) for all $C \in \mathcal{C}$ and for all $x \in V(C)$ it holds:

b1) $OCC(x, C_+) \leq 1$

b2) $\tau_{max}(x, C_+) \leq \tau_{min}(x, C_-)$.

Like KI, D_2 can be decided by positive hyperresolution; note that KI, D_2 are not comparable (none is contained in the other one) because in KI the restriction on the facts is stronger, for D_2 the restriction b1) also applies to rules and b2) is more restricted. D_2 does not contain DATALOG, while KI does.

Class D_3:

A set of clauses $\mathcal{C}$ belongs to D_3 if

D_3 - a) $\mathcal{C}$ is Horn.

D_3 - b) For all $C \in \mathcal{C}$ and for all $x \in V(C)$ it holds:

b1) $occ(x, C_-) \leq 1$

b2) $\tau_{max}(x, C_-) \leq \tau_{min}(x, C_+)$.

Like in the case of KI, KII, D_3 is the mirror image of D_2. But for D_3 semantic clash resolution with positive setting (= negative hyperresolution) alone does not terminate. Condensing was necessary in order to achieve this effect. Like for KI, KII, PVD we will show now that the Horn structure is not essential for the decidability of D_2, D_3.

<u>DEFINTION</u> 3.2.2.: A set of clauses $\mathcal{C}$ belongs to OCC1N (occurrence in negative part only once) if there exists a setting for $\mathcal{C}$ with the following properties:

OCC1N - 1) $OCC(x, C_{neg}) \leq 1$ for all $C \in \mathcal{C}$, $x \in V(C_{neg})$

OCC1N - 2) $\tau_{max}(x, C_{neg}) \leq \tau_{min}(x, C_{pos})$ for all $C \in \mathcal{C}$ and $x \in V(C_{neg}) \cap V(C_{pos})$.

Note that like in the case PVD, OCC1N generalizes D_2, D_3 also on Horn forms. The proof that RSCmD decides OCC1N, is substantially more technical than this for PVD. Before we start the proof of the main result we first need some technical lemmas concerning the behaviour of m.g.u.'s in resolution and factoring.

<u>DEFINITION</u> 3.2.3.: Let P be an atom considered as a string of symbols. Then symb(i, P) denotes the i-th symbol in P; string(i, P) denotes the string of symbols in P beginning with the i-th symbol.

<u>DEFINITION</u> 3.2.4.: A pair of atoms (P, Q) fulfils (∗) if there is a number k s.t. it holds:

(∗1) $symb(i, P) = symb(i, Q)$ for all $1 \leq i < k$.

(∗2) $OCC(x, string(k, Q)) = 1$ for all $x \in V(string(k, Q))$ and $V(P) \cap V(string(k, Q)) = \emptyset$.

The property (∗) occurs in a natural way if we unify {P, Q} where P is an arbitrary atom and $OCC(x, Q) = 1$ for all $x \in V(Q)$.

As an example, let $R = P(x, f(x), y)$, $Q = P(f(x_1), x_2, x_3)$.
Setting k = 3 we get P(, P(as strings of the first two symbols from R and Q respectively and thus (∗1), and $OCC(x_i, f(x_1), x_2, x_3) = 1$ for $x_i \in \{x_1, x_2, x_3\}$ and $V(f(x_1), x_2, x_3) \cap V(R) = \emptyset$. Computing the first mesh substituent $\{f(x_1)/x\}$ of the m.g.u. for {P, Q} we get $R' = R(f(x_1), f(f(x_1)), y)$ $Q' = Q = P(f(x_1), x_2, x_3)$. Now (∗) holds for (P', Q') with k = 8. Note that the number k depends on the syntax representation of atoms, but not the property (∗) itself. The length (= number of symbols) of an atom does not coincide with the corresponding notion usual in complexity theory, where only a finite alphabet is allowed. Here we work over the infinite alphabet of predicate logic.

We are now in the position to prove a structural property of most general unifiers unifying {P, Q} where (P, Q) fulfils (∗):

<u>LEMMA</u> 3.2.1.: Suppose that (P, Q) fulfils (∗) and that {P, Q} is unifiable. Then the m.g.u. of {P, Q} is of the form $\lambda \cup \mu$, where $dom(\lambda) \subseteq V(P)$, $\lambda = \{t_1/x_1, \ldots, t_k/x_k\}$ where all t_i are subterms of Q and $dom(\mu) \subseteq V(Q) - V(P)$.

<u>Remark:</u>

Note that, in general, m.g.u.'s are not of the form expressed in lemma 3.2.1. Just take $P = P(x, f(x), f(y))$ and $Q = P(f(u), v, u)$; then the m.g.u. is

$$\sigma = \{ f^{(2)}(y)/x,\ f^{(3)}(y)/v,\ f(y)/u \}.$$

Obviously, σ cannot be represented as $\lambda \cup \mu$ in the sense of lemma 3.2.1.

<u>Proof of lemma 3.2.1.:</u>

By induction on $n = |V(\{P, Q\})|$.

$n = 0$: $\{P, Q\}$ is only unifiable if $P = Q$ and therefore the m.g.u. is $\emptyset$.
We only have to set $\lambda = \mu = \emptyset$.

(IH) Suppose that for all (P, Q) fulfilling $(\bullet)$ and $|V(\{P, Q\})| = n$ the m.g.u. is of the form $\lambda \cup \mu$ as indicated above.

<u>Case n + 1:</u>

If $P = Q$ then $\lambda = \mu = \emptyset$ and the assertion holds.
Otherwise the disagreement set $D(P, Q)$ is $\{x, t\}$ for a variable x and a term t. Note that we unify from left two right and (x, t) is the first pair of different corresponding terms for (P, Q).
By Robinson's unification algorithm we get
$\text{m.g.u.}(P, Q) = \{t/x\}\ \text{m.g.u.}(P\{t/x\}, Q\{t/x\})$.
But because $x \notin V(t)$ we get $|V(\{P\{t/x\}, Q\{t/x\})| = n$.
In order to apply (IH) we have to show first that $(P\{t/x\}, Q\{t/x\})$ fulfils $(\bullet)$.

Case a) $x \in V(P)$.

If $x = \text{symb}(k, P)$ then by definition of the disagreement set $\text{symb}(i, P) = \text{symb}(i, Q)$ for $i < k$ and k is the maximal number with this property.
Because (P, Q) fulfils $(\bullet)$ we get that $OCC(x, \text{string}(Q, k)) = 1$ for all $x \in V(\text{string}(Q, k))$.
Because $x \in V(P)$ we get $Q\{t/x\} = Q$ and $(P\{t/x\}, Q\{t/x\})$ fulfils $(\bullet)$ for the same number k.

case b) $x \in V(Q)$.

Let m be the maximal number s.t. $\text{symb}(i, P\{t/x\}) = \text{symb}(i, P) = \text{symb}(i, Q\{t/x\})$. Obviously $m > k$ holds.
Because $OCC(x, \text{string}(Q,k)) = 1$ and $x = \text{symb}(k, Q)$ we get $OCC(y, \text{string}(Q\{t/x\}, m+1)) = 1$ for all y in $V(\text{string}(Q\{t/x\}, m+1))$.
Thus $(\bullet)$ holds for $(P\{t/x\}, Q\{t/x\})$ by setting k to $m+1$.

Thus we know that $(P\{t/x\}, Q\{t/x\})$ fulfils $(\ast)$ in any case and we turn to $\sigma = \text{m.g.u.}(\{P\{t/x\}, Q\{t/x\}\})$.

By (IH) $\sigma = \{t_1/x_1\} \cup \ldots \cup \{t_k/x_k\} \cup \mu$ where $\{x_1, \ldots, x_k\} \subseteq V(P\{t/x\})$, $\text{dom}(\mu) \subseteq V(Q\{t/x\}) - V(P\{t/x\})$ and $t_1, \ldots, t_k$ are subterms of Q.

1) If $x \in V(P)$ then t is a term in Q and we get
$\text{m.g.u.}(P, Q) = \{t/x\} (\{t_1/x_1\} \cup \ldots \cup \{t_k/x_k\} \cup \mu)$.
Because (P, Q) fulfils $(\ast)$ by position k (see case a) above), we know that $V(P) \cap V(\text{string}(k, Q)) = \emptyset$. But t is a prefix of string(k, Q) and thus $V(P) \cap V(t) = \emptyset$. Particularly $x_i \notin V(t)$ for $i = 1, \ldots, k$.
By $\text{dom}(\mu) \subseteq V(Q\{t/x\}) - V(P\{t/x\})$ and $V(t) \subseteq V(P\{t/x\})$ we also have $\text{dom}(\mu) \cap V(t) = \emptyset$.

Consequently we get $\text{m.g.u.}(P, Q) = \{t/x\} \cup \{t_1/x_1\} \cup \ldots \cup \{t_k/x_k\} \cup \mu$.
Moreover we have $\{x, x_1, \ldots, x_k\} \subseteq V(P)$.
It remains to show $\text{dom}(\mu) \subseteq V(Q) - V(P)$. Because $\{x\} = V(P) - V(P\{t/x\})$ it is enough to show that $x \notin \text{dom}(\mu)$; the latter property is obvious, because $x \notin V(\{P\{t/x\}, Q\{t/x\}\})$.
This settles the case n+1 for $x \in V(P)$.

2) $x \in V(Q)$ and t is a term in P.
Then
$$\text{m.g.u}\{P,Q\} = \{t/x\}(\{t_1/x_1\} \cup \ldots \cup \{t_k/x_k\} \cup \mu) = \{t_1/x_1\} \cup \ldots \cup \{t_k/x_k\} \cup \{t'/x\}\mu$$
for some term t'.
Set $\lambda = \{t_1/x_1\} \cup \ldots \cup \{t_k/x_k\}$. We have to show that $\{x_1, .. x_k\} \subseteq V(P)$ holds; but $V(P) = V(P\{t/x\})$ and $\{x_1, \ldots x_k\} \subseteq V(P\{t/x\})$ by (IH).
Thus $\text{dom}(\lambda) \subseteq V(P)$, $\lambda = \{t_1/x_1\} \cup \ldots \cup \{t_k/x_k\}$ for subterms t_i in Q.

It remains to show that $\text{dom}(\mu') \subseteq V(Q) - V(P)$ for $\mu' = \{t'/x\}\mu$.
But
$$\text{dom}(\mu) \subseteq V(Q\{t/x\}) - V(\{P\{t/x\}) = V(Q\{t/x\}) - V(P) \subseteq V(Q) - V(P).$$

By $\text{dom}(\mu') \subseteq \{x\} \cup \text{dom}(\mu)$ and by $x \in V(Q)$ we conclude $\text{dom}(\mu') \subseteq V(Q) - V(P)$. This concludes the case $n + 1$ for $x \in V(Q)$.

Q.E.D.

The following lemma shows that factoring of negative clauses in OCC1N preserves the property OCC1N and does not increase term depth.

LEMMA 3.2.2. Let $C \in \mathcal{C}$ for $\mathcal{C} \in$ OCC1N w.r.t. to a setting $\mathfrak{m}$ s.t. C is false in $\mathfrak{m}$ and let $C\vartheta$ be a factor of C. Then $\{C\vartheta\} \in$ OCC1N and $\tau(C\vartheta) \leq \tau(C)$.

Proof:
By definition of OCC1N we have $OCC(x, C) = 1$ for all $x \in V(C)$.
Furthermore we note that every factor can be computed by iterated binary factoring.
Because $\mathcal{C} \in$ OCC1N we also have $V(L) \cap V(M) = \emptyset$ for $L \neq M$ and $L, M \in C$.
If a binary factoring substitution ϑ unifies $\{L, M\}$ for $L, M \in C$ we get:

$$C\vartheta = (C - \{L, M\}) \cup \{L, M\}\vartheta.$$

Because $\tau(C - \{L, M\}) \leq \tau(C)$ anyway, it suffices to show that
$\{L\vartheta\} \in$ OCC1N and
$\tau(L\vartheta) \leq \max\{\tau(L), \tau(M)\}$ $(\leq \tau(C))$.

Let P,Q be the corresponding atom formulas to L, M.
Then (P, Q) fulfils condition $(\bullet)$ and by lemma 3.2.1 we have
$\vartheta = \{t_1/x_1, \ldots, t_n/x_n\} \cup \mu$ for $\{x_1, \ldots, x_n\} \subseteq V(P)$ $db(\mu) \subseteq V(Q)$ and the t_i are subterms of Q.
Therefore we have $P\vartheta = P\{t_1/x_1, \ldots, t_n/x_n\}$.
But by $\{C\} \in$ OCC1N we know that $OCC(x, t_i) = 1$ for all $x \in V(t_i)$ and $i = 1, \ldots, n$; moreover we have $V(t_i) \cap V(t_j) = \emptyset$ for all $i, j \in \{1, \ldots n\}$ s.t. $i \neq j$.
Thus we get $OCC(x, P\vartheta) = 1$ for $x \in V(P\vartheta)$.
Because also (Q, P) fulfils $(\bullet)$ we conclude $OCC(x, Q\vartheta) = 1$ for all $x \in V(Q\vartheta)$ in the same way.
Because ϑ is a m.g.u. we also get $V(C - \{L, M\}) \cap V(L\vartheta) = \emptyset$ and by

$$C - \{L, M\} \in \text{OCC1N}, \quad (C - \{L, M\}) \cup \{L\vartheta\} \in \text{OCC1N}.$$

It remains to show $\tau(N\vartheta) \leq \max\{\tau(P), \tau(Q)\}$ for $N = P, Q$.

Again, let $\vartheta = \{ t_1/x_1, \ldots, t_n/x_n\} \cup \mu$ and

$\eta = \{ t_1/x_1, \ldots, t_n/x_n\}$.

We get $P\vartheta = Q\vartheta$ and also $P\eta = Q\mu$:

We first show $\tau(P\vartheta) = \tau(P\eta) \leq \max\{\tau(P), \tau(Q)\}$.
Because every x_i occurs only once in P we get

$$\tau(P\{t_i/x_i\}) \leq \max\{\tau(P), \tau(Q)\} \quad \text{for every } i = 1, \ldots, n;$$

$t_i/x_i \in \vartheta$ means that (x_i, t_i) is a corresponding term pair in (P, Q) and thus $\tau_{max}(x_i, P) = \tau(x_i, P) = \tau(t_i, Q)$. Because x_i is only substituted at the place of the corresponding pair we get

$$\tau(P\{t_i/x_i\}) \leq \max\{\tau(P), \tau(Q)\}$$

(if $\tau(P) \leq \tau(Q)$ in advance then $\tau(P\{t_i/x_i\}) \leq \tau(Q)$).

But either $\tau(P\eta) = \tau(P)$ and thus $\tau(P\eta) \leq \max\{\tau(P), \tau(Q)\}$ or

$\tau(P\eta) = \tau(x_i, P) + \tau(t_i)$ for some $i \in \{1, \ldots n\}$
(note that by $OCC(x_i, P) = 1$ $\tau_{max}(x_i, P) = \tau(x_i, P)$). But

$$\tau(x_i, P) + \tau(t_i) = \tau(P\{t_i/x_i\}) \leq \max\{\tau(P), \tau(Q)\}$$

by the argumentation above.

It follows

$$\tau(P\vartheta) = \tau(P\eta) \leq \max\{\tau(P), \tau(Q)\}.$$

But $Q\vartheta = P\vartheta$ and thus also

$$\tau(Q\vartheta) \leq \max\{\tau(P), \tau(Q)\}.$$

We get $\tau(C\vartheta) = \tau((C - \{L, M\}) \cup \{P\vartheta\})$ and by
$\tau(\{L, M\}\vartheta) = \tau(L\vartheta) = \tau(P\vartheta) \leq \max\{\tau(P), \tau(Q)\}$
$\tau(C\vartheta) = \tau(C)$.

Q.E.D.

In the next lemma we show that OCC1N is closed under binary semantic resolution and that the term depth of the resolvents cannot increase.

<u>LEMMA</u> 3.2.3.: Let $\{C, D\} \in$ OCC1N w.r.t. a setting $\mathfrak{m}$; let C be positive in $\mathfrak{m}$, D be negative in $\mathfrak{m}$ and R be a binary $\mathfrak{m}$-resolvent of C and D. Then $\{R\} \in$ OCC1N and $\tau(R_{neg}) \leq \max\{\tau(C_{neg}), \tau(D)\}$.

<u>Proof</u>:

Let $C = C_{neg} \cup C_{pos}$, $L \in C_{pos}$ and $M \in D$ s.t. ϑ is a m.g.u. of $\{L, M^d\}$ and $R = C_{neg}\vartheta \cup (D - \{M\})\vartheta \cup (C_{pos} - \{L\})\vartheta$ is a binary semantic resolvent of C and D.

Because $R_{neg} = C_{neg}\vartheta \cup (D - \{M\})\vartheta$ we have to show

$$\tau(C_{neg}\vartheta \cap (D - \{M\})\vartheta) \leq \max\{\tau(C_{neg}), \tau(D)\}.$$

Because $\{D\} \in$ OCC1N, it is variable - decomposed; because ϑ is a m.g.u. we have $dom(\vartheta) \cap V(D - \{M\}) = \emptyset$ and thus $(D - \{M\})\vartheta = (D - \{M\})$. Because, trivially, $\tau(D - \{M\}) \leq \tau(D)$ it suffices to show $\tau(C_{neg}\vartheta) \leq \max\{\tau(C_{neg}), \tau(D)\}$. We further reduce the unequality to be proved to

$$\tau(C_{neg}\vartheta) \leq \max\{\tau(C_{neg}), \tau(M)\}.$$

Because M is a literal from a false clause, we get that (P, Q) fulfils $(\bullet)$ for $P = at(L)$, $Q = at(M)$.

By lemma 3.2.1. we know that $\vartheta = \{t_1/x_1, \dots, t_n/x_n\} \cup \mu$ where $\{x_1, \dots, x_n\} \subseteq V(P)$, t_i are subterms of Q and $dom(\mu) \subseteq V(Q)$.
It follows that $C_{neg}\vartheta = C_{neg}\sigma$ for $\sigma = \{t_1/x_1, \dots, t_n/x_n\}$.

If $\tau(C_{neg}\sigma) = \tau(C_{neg})$ then $\tau(C_{neg}\vartheta) = \tau(C_{neg}\sigma) \leq \max\{\tau(C_{neg}), \tau(M)\}$ and we have what we want.

If $\tau(C_{neg}\sigma) > \tau(C_{neg})$ then there is an $i \in \{1, \dots, n\}$ s.t.

(A) $\tau(C_{neg}\sigma) = \tau_{max}(x_i, C_{neg}) + \tau(t_i)$ for an $x_i \in V(C_{neg}) \cap V(C_{pos})$.
But t_i is a term which occurs in M and therefore

(B) $\tau(t_i) \leq \tau(M) - \tau_{min}(x_i, L)$
(note that, by $t_i/x_i \in \lambda$, (x_i, t_i) is a corresponding pair in (P, Q)).

But by condition OCC 1N) - 2) and $x_i \in V(C_{neg}) \cap V(C_{pos})$ we get

$$\tau_{max}(x_i, C_{neg}) \leq \tau_{min}(x_i, C_{pos})$$

and also $\tau_{min}(x_i, C_{pos}) \leq \tau_{min}(x_i, L)$.

Therefore: (C) $\tau_{max}(x_i, C_{neg}) \leq \tau_{min}(x_i, L)$.

Combining (A) and (C) we get the inequality

$$\tau(C_{neg}\sigma) \leq \tau_{min}(x_i, L) + \tau(t_i)$$

and using (B):

$$\tau(C_{neg}\sigma) \leq \tau_{min}(x_i, L) + \tau(M) - \tau_{min}(x_i, L) = \tau(M).$$

Thus, in any case $\tau(C_{neg}\sigma) \leq \max\{\tau(C_{neg}), \tau(M)\}$ and we have proved

$$\tau(R_{neg}) \leq \max\{\tau(C_{neg}), \tau(D)\}.$$

It remains to show that $\{R_{neg}\} \in$ OCC1N holds.

For this purpose we have to show first that $OCC(x, R_{neg}) = 1$ for all $x \in R_{neg}$.

But $R_{neg} = C_{neg}\vartheta \cup (D - \{M\})$, where $D - \{M\} \in$ OCC1N and
$V(C_{neg}\vartheta) \cap V(D - \{M\}) = \emptyset$.

Therefore it is sufficient to prove $OCC(x, C_{neg}\vartheta) = 1$ for all $x \in V(C_{neg}\vartheta)$.

But $C_{neg}\vartheta = C_{neg}\sigma$ for $\sigma = \{t_1/x_1, \dots, t_n/x_n\}$ where $OCC(x, t_i) = 1$ for all $x \in V(t_i)$, $i \in \{1, \dots, n\}$ and $V(t_i) \cap V(t_j) = \emptyset$ for $i, j \in \{1, \dots, n\}$ and $i \neq j$.

It follows immediately that, by $C \in$ OCC1N, $OCC(x, C_{neg}\sigma) = 1$ for all $x \in V(C_{neg}\sigma)$.

We get that OCC1N - 1) holds for R_{neg}.

To show OCC1N - 2) we have to prove that

$$\tau_{max}(x, R_{neg}) \leq \tau_{min}(x, R_{pos}) \text{ for all } x \in V(R_{neg}) \cap V(R_{pos}).$$

But $R_{neg} = C_{neg}\vartheta \cup (D - \{M\})$ where $V(D - \{M\}) \cap V(R_{pos}) =$
$= V(D - \{M\}) \cap V((C_{pos} - \{L\})\vartheta) = \emptyset$

(note that $V(C) \cap V(D) = \emptyset$ before resolution and $dom(\vartheta) \cap V(D - \{M\}) = \emptyset$).

Thus it suffices to prove

$$\tau_{max}(x, C_{neg}\vartheta) \leq \tau_{min}(x, (C_{pos} - \{L\})\vartheta) \quad \text{for all}$$
$$x \in V(C_{neg}\vartheta) \cap V((C_{pos} - \{L\})\vartheta).$$

Suppose now that $x \in V(C_{neg}\vartheta) \cap V((C_{pos} - \{L\})\vartheta)$.
If $x \in V(C_{neg}) \cap V((C_{pos} - \{L\}))$ and $x \notin dom(\vartheta)$ then
$x \in V(C_{neg}\vartheta) \cap V((C_{pos} - \{L\})\vartheta)$ and

a) $\tau_{max}(x, C_{neg}\vartheta) = \tau_{max}(x, C_{neg})$,
b) $\tau_{min}(x, (C_{pos} - \{L\})\vartheta) = \tau_{min}(x, C_{pos} - \{L\})$.

Note that $C\vartheta = C\sigma$ for $\sigma = \{t_1/x_1, \dots, t_n/x_n\}$, where the t_i are terms in M. Because $V(t_i) \cap V(C) = \emptyset$ for all i, the variable x cannot occur in $rg(\sigma)$. We thus have a guarantee that x is not introduced additionally via ϑ and the occurrences of x in $C_{neg}\vartheta$ are those of C_{neg}; this yields the properties (a) and (b).

By OCC1N - 2 for C we conclude

$$\tau_{max}(x, C_{neg}\vartheta) = \tau_{max}(x, C_{neg}) \leq \tau_{min}(x, C_{pos}) \leq \tau_{min}(x, C_{pos} - \{L\}) = \\ = \tau_{min}(x, (C_{pos} - \{L\})\vartheta).$$

If $x \in V(C_{neg}) \cap V(C_{pos} - \{L\})$ and $x \in dom(\vartheta)$ then, as m.g.u.'s are idempotent, $x \notin V(C_{neg}\vartheta) \cap V((C_{pos} - \{L\})\vartheta)$.
If $x \notin V(C_{neg}) \cap V(C_{pos} - \{L\})$ then x must occur in $rg(\vartheta)$.
Because $C_{neg}\vartheta = C_{neg}\sigma$ for $\sigma = \{t_1/x_1, \dots, t_n/x_n\}$ and $dom(\sigma) \subseteq V(L)$, x must occur in some of the t_i. Because the t_i are mutually variable - disjoint, x can only occur in one of the t_i's.

Thus suppose $x \in V(t_i)$.
Because $x \notin V(C_{neg}) \cap V(C_{pos} - \{L\})$ we get

(1) $\tau_{max}(x, C_{neg}\vartheta) = \tau_{max}(x_i, C_{neg}) + \tau_{max}(x, t_i)$ and
(2) $\tau_{min}(x, (C_{pos} - \{L\})\vartheta) = \tau_{min}(x_i, C_{pos} - \{L\}) + \tau_{min}(x, t_i)$.

But as t_i is a term occuring in M, $OCC(x, t_i) = 1$ and

(3) $\tau_{max}(x, t_i) = \tau_{min}(x, t_i)$

By OCC1N - 2) we know that

$$\tau_{max}(x_i, C_{neg}) \leq \tau_{min}(x_i, C_{pos})$$

and therefore

(4) $\tau_{max}(x_i, C_{neg}) \leq \tau_{min}(x_i, C_{pos} - \{L\})$

combining (1) - (4) we easily get

$$\tau_{max}(x, C_{neg}\vartheta) \leq \tau_{min}(x, (C_{pos} - \{L\})\vartheta).$$

Q.E.D.

We are now in the position to show that OCC1N is closed under SC$\mathfrak{m}$ - resolution and that the term depth of clash resolvents cannot increase.

<u>LEMMA</u> 3.2.4.: Let $\mathcal{C}$ be a set of clauses in OCC1N w.r.t. the setting $\mathfrak{m}$ and $\Gamma = (C; D_1, \ldots, D_n)$ be a semantic clash of clauses in OCC1N. If R is a clash resolvent of Γ then $\{R\} \in$ OCC1N and $\tau(R) \leq \max\{\tau(C), \tau(D_1), \ldots, \tau(D_n)\}$.

<u>Proof</u>:

If R is a clash resolvent of Γ then R is R_n for some R_n defined as:

$R_0 = C$

R_{i+1} = a binary semantic resolvent of R_i and (a factor of) D_{i+1} for $i < n$.

We prove by induction on i that $\tau((R_i)_{neg}) \leq \max\{\tau(C), \tau(D_1), \ldots, \tau(D_n)\}$ and $\{R_i\} \in$ OCC1N.

$i = 0$: $R_0 = C$ and $\{C\} \in$ OCC1N; moreover, $\tau(C_{neg}) \leq \tau(C)$.

Suppose that the assertion holds for i.

If $i = n$ we are done.

Thus assume that $i < n$.

Then R_{i+1} is a binary resolvent of R_i and D_{i+1}', where $D_{i+1}' = D_{i+1}$ or $D_{i+1}' = D_{i+1}\eta$ for a factoring substitution η.

Because $\{D_{i+1}\} \in$ OCC1N , we know by lemma 3.2.2. that $D_{i+1}' \in$ OCC1N and $\tau(D_{i+1}') = \tau(D_{i+1})$.

By lemma 3.2.3. we get

$\tau((R_{i+1})_{neg}) \leq \max\{\tau((R_i)_{neg}), \tau(D_{i+1}')\} \leq \max\{\tau(R_i)_{neg}, \tau(D_{i+1})\}$ and $\{R_{i+1}\} \in$ OCC1N.

Applying the induction hypothesis we thus obtain

$\tau((R_{i+1})_{neg}) \leq \max\{\tau(C), \tau(D_1), \ldots, \tau(D_n)\}$.

But because R_n is a clash resolvent we also have $(R_n)_{neg} = R_n$ and therefore

$\tau(R) = \tau(R_n) \leq \max\{\tau(C), \tau(D_1), \ldots, \tau(D_n)\}$.

Q.E.D.

<u>THEOREM</u> 3.2.2.: **SC$\mathfrak{m}$D - resolution is a decision procedure for OCC1N. More precisely:**
If $\mathcal{C} \in$ OCC1N then there is a setting $\mathfrak{m}$ s.t. $R^{*}_{\mathfrak{m}}(\mathcal{C})$ is finite (we take this $\mathfrak{m}$ which fulfils OCC1N)-1) OCC1N)-2)). Moreover there is an algorithm defining for every $\mathcal{C} \in$ OCC1N a setting $\mathfrak{m}$ s.t. $R^{*}_{\mathfrak{m}}(\mathcal{C})$ is finite.

<u>Proof</u>:
We first show that $R^{*}_{\mathfrak{m}}(\mathcal{C}) \in$ OCC1N for $\mathfrak{m}$ s.t. $\mathcal{C} \in$ OCC1N w.r.t. $\mathfrak{m}$.
Because $\bigcup_{i=0}^{\infty} R^{i}_{\mathfrak{m}}(\mathcal{C}) = R^{*}_{\mathfrak{m}}(\mathcal{C})$ we prove $R^{i}_{\mathfrak{m}}(\mathcal{C}) \in$ OCC1N by induction on i.

$R^{0}_{\mathfrak{m}}(\mathcal{C}) = \mathcal{C}_0 = \{C \mid C \in \mathcal{C}, C \text{ true in } \mathfrak{m}\} \cup \{C_{cond} \mid C \in \mathcal{C}, \mathcal{C} \text{ false in } \mathfrak{m}\}$.
Because for every false C, OCC1N)-1) holds, also for the condensation (because it is valid for any factoring) we get $\mathcal{C}_0 \in$ OCC1N.
(IH) Suppose that $R^{i}_{\mathfrak{m}}(\mathcal{C}) \in$ OCC1N.

Then $R^{i+1}_{\mathfrak{m}}(\mathcal{C}) = (R^{i}_{\mathfrak{m}}(\mathcal{C}) \cup \text{RSC}\mathfrak{m}\text{D}(R^{i}_{\mathfrak{m}}(\mathcal{C})))/_{\sim_V}$.

By (IH) and lemma 3.2.4. we know that $\text{RSC}\mathfrak{m}(R^{i}_{\mathfrak{m}}(\mathcal{C})) \in$ OCC1N.
But every clause $C \in \text{RSC}\mathfrak{m}(R^{i}_{\mathfrak{m}}(\mathcal{C}))$ is false and in OCC1N.

By lemma 3.2.2., every factor of C is in OCC1N; we conclude that also $C_{cond} \in$ OCC1N; thus $\text{RSC}\mathfrak{m}\text{D}(R^{i}_{\mathfrak{m}}(\mathcal{C})) \in$ OCC1N.
So we get $R^{i+1}_{\mathfrak{m}}(\mathcal{C}) \in$ OCC1N and, by induction, $R^{*}_{\mathfrak{m}}(\mathcal{C}) \in$ OCC1N.

It remains to show, that $R^{*}_{\mathfrak{m}}(\mathcal{C})$ is finite:
By an induction argument analogous to that before we conclude $\tau(R^{i}_{\mathfrak{m}}(\mathcal{C})) \leq \tau(\mathcal{C})$ for all i (we just use lemma 3.2.4 and the fact that condensed clauses cannot be of higher depth than uncondensed ones).
But $R^{*}_{\mathfrak{m}}(\mathcal{C}) - \mathcal{C}$ consists of negative clauses only. By definition of OCC1N, every $C \in R^{*}_{\mathfrak{m}}$ is decomposed (even after condensation). But there can be only finitely many decomposed, condensed clauses of finite fixed depth (note that there may be no literals which are variants of other literals in C).
It follows that $R^{*}_{\mathfrak{m}}(\mathcal{C})$ is finite.

Finally we observe that $\mathfrak{m}$ can be found effectively, as in the case PVD: There are only finitely many settings possible for $\mathcal{C}$ and OCC1N) -1), OCC1N) - 2) is decidable for every $\mathfrak{m}$.

Q.E.D.

$SC\mathfrak{m}D$ - resolution can be used as decision method for both PVD and OCC1N (thus for PVD $\cup$ OCC1N). If $\mathcal{C} \in$ PVD $\cup$ OCC1N we find some setting $\mathfrak{m}$ s.t. $\mathcal{C} \in$ PVD w.r.t. $\mathfrak{m}$ or $\mathcal{C} \in$ OCC1N w.r.t. $\mathfrak{m}$ (possibly in both classes) and may compute $R^{*}_{\mathfrak{m}}(\mathcal{C})$.

Rather than just a decision "procedure" it is a decision method by which the resolution procedure corresponding to a setting $\mathfrak{m}$ can be found algorithmically. This approach is somewhat different from Joyner's, where a single refinement is specified to decide some classes. But similarly we only define complete resolution refinements, which at the same time can be used as ordinary theorem provers. Even if $\mathcal{C} \notin$ PVD $\cup$ OCC1N we may try to find a setting s.t. as many clauses as possible fulfil the conditions PVD1), PVD2) or OCC1N - 1), OCC1N) - 2). Although we can no longer guarantee termination, the procedure favours the production of clauses having low term depth. In this sense it can provide a method to control growth of nesting based on the logical structure of the set of input clauses rather than a method based on heuristics only.

In case of the Bernays - Schönfinkel class (BS), the class PVD can be used to obtain an efficient decision algorithm.

Let BS = $\{(\exists \bar{x})(\forall \bar{y}) M(\bar{x}, \bar{y}) \mid M(\bar{x}, \bar{y})$ is a function - free matrix with variable vectors $\bar{x}, \bar{y}\}$.

Then the clause form of BS is $BS^{*} = \{\mathcal{C} \mid \tau(\mathcal{C}) = 0\}$.

BS^{*} is easily decidable by a ground method because for every $\mathcal{C} \in BS^{*}$ the Herbrand universe is finite.

A straight forward decision procedure is the following:

1) Compute all ground instances of clauses in $\mathcal{C}$.

2) Apply the Davis - Putnam procedure [DP 60].

This method, although easily defined, can be very expensive because the number of ground instances can be very high (for every clause C we obtain $|H(\mathcal{C})|^{|V(c)|}$ ground instances).

Thus resolution on $\mathcal{C}$ itself might be much more promising. But unfortunately BS^{*} cannot be decided by semantic clash resolution at all.

<u>EXAMPLE</u> 3.2.3.:

$\mathcal{C} = \{\{P(x,z,u),\ \neg P(x,y,u),\ \neg P(y,z,u)\},$

$\{P(x,x,a)\},$

$\{\neg P(x,z,u), P(x,y,u),\ P(y,z,u)\}$

$\{\neg P(x,x,b)\}\}$ (a, b are different constant symbols).

As setting we only have negative setting ($\mathfrak{m}_n$) and positive setting ($\mathfrak{m}_p$). Under RSC$\mathfrak{m}_p$D we get an infinite sequence of clauses generated by the first and the fourth clause, which cannot be eliminated by condensing or by subsumption. For RSC$\mathfrak{m}_n$D a similar thing happens with the second and the third clause. This proves that no semantic refinement can decide BS*. Obviously neither $\mathfrak{m}_p$ nor $\mathfrak{m}_n$ fulfils conditions PVD1), PVD2) and therefore $\mathcal{C} \notin$ PVD.
However we will define a simple method to transform a set of clauses $\mathcal{C}$ in BS* into a set of clauses $\mathcal{C}' \in$ PVD s.t. $\mathcal{C}$ and $\mathcal{C}'$ are sat-equivalent. We proceed as follows:

case a) $\mathcal{C} \in$ PVD: $\mathcal{C}' = \mathcal{C}$.

case b) $\mathcal{C} \notin$ PVD:

b1) choose a setting $\mathfrak{m}$ for $\mathcal{C}$.

b2) For all $C \in \mathcal{C}$ we define:

If $V(C_{neg}) \subseteq V(C_{pos})$ then $T(C) = \{C\}$.

If $V(C_{neg}) - V(C_{pos}) \neq \emptyset$ then

$T(C) = \{C\lambda \mid dom(\lambda) = V(C_{neg}) - V(C_{pos}),\ rg(\lambda) \subseteq H(\mathcal{C})\}$.

b3) $\mathcal{C}' = \bigcup_{C \in \mathcal{C}} T(C)$

In both cases a), b) $\mathcal{C}' \in$ PVD and $\mathcal{C}'$ is sat-equivalent to $\mathcal{C}$.
After definition of $\mathcal{C}'$ we only need to compute $R^*_{\mathfrak{m}}(\mathcal{C}')$.
The algorithm above can be optimized by searching for an $\mathfrak{m}$ s.t. $\mathcal{C}'$ becomes minimal (this can be done without actually computing $\mathcal{C}'$). In general the size of $\mathcal{C}'$ will be much smaller than the set of all ground instances of $\mathcal{C}$; in such a case the method above is clearly superior to the ground method. We return to example 3.2.3.:

$\mathcal{C} \notin$ PVD, so we choose $\mathfrak{m}_p$ and compute $\mathcal{C}'$.

$\mathcal{C}' = \{\{P(x,z,u),\ \neg P(x,a,u),\ \neg P(a,z,u)\},$

$\{ P(x,z,u),\ \neg P(x,b,u),\ \neg P(b,z,u) \},$

$\{ P(x,x,a) \},$

$\{ \neg P(x,z,u),\ P(x,y,u),\ P(y,z,u) \},$

$\{ \neg P(a,a,b)\},\ \{ \neg P(b,b,b) \} \}.$

$|\mathcal{C}'| = |\mathcal{C}| + 2$ and $\mathcal{C}' \in$ PVD.

For practical purposes we may apply subsumption; because $SC\mathfrak{m}$-resolution + subsumption is complete and $RSC\mathfrak{m}$ decides PVD, so $RSC\mathfrak{m}$ + subsumption also does. Let sub($\mathcal{C}$) the subset of $\mathcal{C}$ obtained by $\mathcal{C}$ after deleting subsumed clauses; instead of $R^{*}_{\mathfrak{m}}$ we define $R^{*}_{\mathfrak{m},s}$ as follows:

$$R^{0}_{\mathfrak{m},s}(\mathcal{C}) = \mathrm{sub}(\mathcal{C}),$$
$$R^{i+1}_{\mathfrak{m},s}(\mathcal{C}) = \mathrm{sub}(R^{i}_{\mathfrak{m},s}(\mathcal{C}) \cup RSC\mathfrak{m}(R^{i}_{\mathfrak{m},s}(\mathcal{C}))),$$
$$R^{*}_{\mathfrak{m},s}(\mathcal{C}) = \bigcup_{i \in N} R^{i}_{\mathfrak{m},s}(\mathcal{C}).$$

By completeness under subsumption $\Box \in R^{*}_{\mathfrak{m},s}(\mathcal{C})$ iff $\Box \in R^{*}_{\mathfrak{m}}(\mathcal{C})$.

We now compute $R^{*}_{\mathfrak{m}_p,s}(\mathcal{C}')$

$\mathrm{sub}(\mathcal{C}') = \mathcal{C}'$

$RSC\mathfrak{m}_p(\mathcal{C}') = \{\{ \neg P(a,a,b)\},\ \{\neg P(b,b,b)\},\ \{ \neg P(a,b,b),\ \neg P(b,a,b)\}\}$ and

$\mathcal{C}_1' = \mathrm{sub}(\mathcal{C}' \cup RSC\mathfrak{m}_p(\mathcal{C}')) = \mathcal{C}' \cup \{\{ \neg P(a,b,b),\ \neg P(b,a,b)\}\}.$

$RSC\mathfrak{m}_p(\mathcal{C}_1') = RSC\mathfrak{m}_p(\mathcal{C}') \cup \{\{\neg P(b,a,b),\ \neg P(a,b,b),\ \neg P(b,b,b)\},$

$\{\neg P(a,b,b),\ \neg P(b,a,b),\ \neg P(a,a,b)\}\}.$

$\mathcal{C}_2' = \mathrm{sub}(\mathcal{C}_1' \cup RSC\mathfrak{m}_p(\mathcal{C}_1')) = \mathcal{C}_1'.$

Therefore $R^{*}_{\mathfrak{m}_p,s}(\mathcal{C}') = \mathcal{C}_1'$ and $\Box \notin \mathcal{C}_1'$; we conclude that $\mathcal{C}'$ (and thus $\mathcal{C}$) is satisfiable.

Using the ground saturation method for the Bernays-Schönfinkel class we obtain a set of 36 ground clauses which have to be tested for satisfiability.

3.3 UNDECIDABLE PROBLEMS

It is natural question, whether the classes PVD and OCC1N are in some sense "arbitrary". First of all PVD contains DATALOG and, like OCC1N, gives a simple clause syntax criterion for decidability. Moreover the classes are quite sharp w.r.t. undecidability. We show this in the case of PVD.

As defined in chapter 3.2. the class KII, defined as: $\mathcal{C} \in$ KII iff

a) $\mathcal{C}$ is Horn,

b) $V(C_-) \subseteq V(C_+)$ and $\tau(C_-) = 0$ for rules $C \in \mathcal{C}$,

c) Goals are ground,

is a subclass of PVD (restriction to positive setting and Horn).

We show now that by introducing the clause $T = \{P(x,z), \neg P(x,y), \neg P(y,z)\}$ (transitivity) into KII we get an undecidable class; while $\tau(T_-) = 0$ we do not have $V(T_-) \subseteq V(T_+)$ (violated by the variable y).

Let KII' = $\{\mathcal{C} \cup \{T\} \,/\, \mathcal{C} \in \text{KII}\}$ ([Lei90]). We show that every equational theory (axiomatized in predicate logic) belongs to KII':

REF) $\{P(x,x)\}$.

SYMM) $\{\neg P(x,y),\ P(y,x)\}$.

T) $\{\neg P(x,y),\ \neg P(y,z),\ P(x,z)\}$.

SUBfi) $\{\neg P(x,y),\ P(f(x_1,\dots,x_{i-1},x,x_{i+1},\dots,x_n),\ f(x_1,\dots,x_{i-1},y,x_{i+1},\dots x_n))\}$
for every function symbol f and every position i in f.

EQ1) $\{P(s_1,t_1)\}$

EQn) $\{P(s_n,t_n)\}$

(EQ1) - EQn) are arbitrary equations; s_i, t_j are terms.

¬CON) $\{\neg P(w_1,w_2)\}$, where w_1, w_2 are ground terms.

All axioms and ¬CON) are in KII', and with exception of T also in KII. The clauses encode the word problem of the equational theory $\{s_1 = t_1, \dots, s_n = t_n\}$, $\neg P(w_1,w_2)$ codes the negation of the equation $v_1 = v_2$ where w_1, w_2 are defined from v_1, v_2 by substituting every variable by a (new) Skolem constant.
It is well known that there are equational theories with undecidable word problems (just take the theory of combinators with η - equality [ST71]).
Thus we see that KII' is undecidable.

It was shown in [Fer91], that the restriction $V(C-) \subseteq V(C+)$ is not only important for Horn - but also for Horn + Krom classes.

Let KIIK be the following class of set of clauses $\mathcal{C}$:

(a) $\mathcal{C}$ is Horn and Krom.

(b) $\tau(C_-) = 0$ for every $C \in \mathcal{C}$.

(c) Goals in $\mathcal{C}$ are ground.

It is obvious, that by adding the condition $V(C-) \subseteq V(C+)$ we get a proper subclass of KII.

We show now that KIIK is an undecidable class.

It was proved in [SS 88], that the class KH2, consisting of all sets of clauses of the form $\mathcal{C} = \{C_1, C_2, F_1, G_1\}$, where C_1, C_2 are two Krom rules, F_1 is a ground fact and G_1 is a ground goal, is undecidable.

To show our result, we construct for every $\mathcal{C} \in$ KH2 a set $\mathcal{C}' \in$ KIIK (by an effective construction) s.t. $\mathcal{C}$ and $\mathcal{C}'$ are sat-equivalent.

The construction works as follows:

Let $C \in \{C_1, C_2\}$ for $\mathcal{C} = \{C_1, C_2, F_1, G_1\}$ as above and let

$$C = \{P(t_1, \dots, t_n), \neg Q(s_1, \dots, s_m)\}.$$

We define

$$T(C) = \Big\{\{P(t_1, \dots, t_n), \neg H_C(x_1, x_1, x_2, x_2, \dots, x_m, x_m, v_1, v_2, \dots, v_k)\},$$
$$\{H_C(s_1, y_1, s_2, y_2, \dots, s_m, y_m, v_1', \dots, v_k'), \neg Q(y_1, \dots, y_m)\}\Big\} \quad \text{where}$$
$$V(C+) \cap V(C-) = \{v_1, \dots, v_k\}.$$

H_C is a new predicate symbol which only occurs in the clauses of $T(C)$.

Observe, that by resolving the two clauses in $T(C)$ we get C back. Thus from $\mathcal{C}' = T(C_1) \cup T(C_2) \cup \{F_1\} \cup \{G_1\}$ we can derive $\mathcal{C}$, and $\mathcal{C}' \in$ KIIK.

It remains to show, that the satisfiability of $\mathcal{C}$ implies that of $\mathcal{C}'$.

Suppose that $\mathcal{C}'$ is unsatisfiable. Then there is a lock refutation of $\mathcal{C}'$. By giving priority to the literals $H_C(\dots)$ and $\neg H_C(\dots)$ in indexing, we get, that all clauses not containing atoms H_{C_1}, H_{C_2} are also derivable from $\{C_1, C_2, F_1, G_1\}$. Therefore $\square \in R^*(\mathcal{C}')$ iff $\square \in R^*(\mathcal{C})$ and $\mathcal{C}, \mathcal{C}'$ are sat-equivalent.

Thus we have shown that KIIK is undecidable.

In KIIK, by the T(C) construction, it is sufficient that two clauses violate $V(C_-) \subseteq V(C_+)$ in order to get undecidability. In KII', by the more general clause form, one clause (it is T) suffices not fulfilling $V(C_-) \subseteq V(C_+)$. Thus it is apparent that KII, and thus also PVD, is quite sharp w.r.t. to the structural property of variable occurrence.

3.4 HORN CLASSES

For the classes in chapter 3. 2. we had to focus on two measures for guaranteeing termination, clause size and term depth.
Restricting the propositional structure of clauses to Horn, we can get rid of the clause size - problem and concentrate on term depth only. The reason is, that under negative setting the only false Horn clauses are unit clauses. Applying hyperresolution w.r.t. $\mathfrak{m}_n$ (negative setting) we only derive positive unit clauses and thus $R^*_{\mathfrak{m}_n}(\mathcal{C})$ is finite if there is a bound on term depth. According to [CL 73] we call $SC\mathfrak{m}_n$ - resolution positive hyperresolution because it produces positive clauses only (in the syntactical sense).
It is almost trivial to verify that the Bernays - Schönfinkel class, restricted to a Horn matrix can be decided by $SC\mathfrak{m}_n$ - resolution. For this purpose we define:

BSH = $\{(\exists\overline{x})(\forall\overline{y})M(\overline{x},\overline{y}) \mid M(\overline{x},\overline{y})$ is a function free matrix in Horn form with free variable vectors $\overline{x},\overline{y}\}$.

By Skolemization we get:

$BSH^* = \{\mathcal{C} \mid \mathcal{C}$ is a set of Horn clauses and $\tau(\mathcal{C}) = 0\}$.

Let $\mathcal{C}$ be in BSH^*. Then $R^*_{\mathfrak{m}_n}(\mathcal{C}) - \mathcal{C}$ consists of positive unit clauses only; because $R^*_{\mathfrak{m}_n}(\mathcal{C})$ does not contain any function symbols, all $C \in R^*_{\mathfrak{m}_n}(\mathcal{C})$ fulfil $\tau(C) = 0$. Let $C \sim_v D$ if D is a variant of C ($C\sigma = D$ for a permutation substitution σ). Clearly $R^*_{\mathfrak{m}_n}(\mathcal{C})/\sim_v$ (the class $R^*_{\mathfrak{m}_n}(\mathcal{C})$ factored under the equivalence relation $\sim_v$) must be finite. Note that we can avoid factoring w.r.t. $\sim_v$ if we keep all clauses in a variable - standard form.
Thus positive hyperresolution decides BSH^* and therefore also DATALOG which is a subclass of BSH^*.

Although we cannot deal with the Bernays-Schönfinkel class using semantic clash resolution only (we refer to the end of chapter 3.2), we can generalize BSH* to classes containing function symbols.
The following class was discussed in [Fer90] (there it was named D_1):

DEFINITION 3.4.1.: VED (variables in equal depth) is the set of all sets of Horn clauses $\mathcal{C}$ s.t. it holds:
For all $C \in \mathcal{C}$ and for all $x \in V(C)$ it holds $\tau_{min}(x, C) = \tau_{max}(x, C)$ (that means, every variable occurs only at a certain fixed depth within a clause).
In BSH* all variables occur at depth 0 and thus trivially BSH* $\subseteq$ VED. Note that in different clauses, the depth of the variables may be different; thus $\{\{\neg P(f(x)), \neg P(g(x))\}, \{\neg Q(x,y), Q(y,x)\}, \{P(a), Q(f(x))\}\} \in$ VED.
We show now, that RSC$\mathfrak{m}_n$ is a decision procedure for VED.
The first step consists in showing that (binary) semantic resolution cannot increase the maximal term depth.

LEMMA 3.4.1.: Let C be a goal or a rule and D be a fact (C is positive in $\mathfrak{m}_n$ and D is negative in $\mathfrak{m}_n$) and let R be a (semantic) resolvent of C, D. Then $\tau(R) \leq \max\{\tau(C), \tau(D)\}$ and
$\{R\} \in$ VED if $\{C, D\} \in$ VED.

Proof:
w.l.o.g. we may assume that C and D are variable disjoint. Let σ be the m.g.u. of the resolvent of C and D. Then σ can be written as concatenation of mesh substituents in the form $\sigma = \{t_1/x_1\} \dots \{t_n/x_n\}$. The substitution t_i/x_i corresponds to the i-th step of the Robinson unification algorithm from left to right.

Let $\varphi_0 = \varepsilon$

$\varphi_{i+1} = \varphi_i\{t_{i+1}/x_{i+1}\}$

Then $\varphi_n = \sigma$ and, by $\tau(R) = \tau((C - M)\sigma) \leq \tau(C\sigma)$ where M is the resolved literal in C, it is sufficient to show by induction on i:

A(i)-1) $\tau(x, C\varphi_i) = \tau(x, D\varphi_i)$ for $x \in V(C\varphi_i) \cap V(D\varphi_i)$

$\tau_{max}(x, C\varphi_i) = \tau_{min}(x, C\varphi_i)$ for $x \in V(C\varphi_i) - V(D\varphi_i)$

$\tau_{max}(x, D\varphi_i) = \tau_{min}(x, D\varphi_i)$ for $x \in V(D\varphi_i) - V(C\varphi_i)$

A(i)-2) $\max\{\tau(C\varphi_i), \tau(D\varphi_i)\} \leq \max\{\tau(C), \tau(D)\}$

A(0) : obvious by $V(C) \cap V(D) = \emptyset$, $\{C, D\} \in$ VED, and by $\varphi_0 = \varepsilon$.

(IH) Suppose that A(i) holds.
Then $\varphi_{i+1} = \varphi_i\{t_{i+1}/x_{i+1}\}$. Because $\{x_{i+1}, t_{i+1}\}$ is a disagreement set for $\{M\varphi_i, P\varphi_i\}$ (for $D = \{P\}$) t_{i+1} is either a term in $M_{\varphi i}$ or in $P\varphi_i$.
Suppose that $SUB(k, M\varphi_i) = x_{i+1}$ and $SUB(k, P\varphi_i) = t_{i+1}$.
By (IH) all variables in t_{i+1} can only occur in some fixed depth and thus also A(i+1)-1) holds for $C\varphi_{i+1}$, $D\varphi_{i+1}$.
The argumentation is the same for $SUB(k, M\varphi_i) = t_{i+1}$, $SUB(k, P\varphi_i) = x_{i+1}$.

ad A(i+1) - 2:
Again suppose w.l.o.g. that $SUB(k, M\varphi_i) = x_{i+1}$ and $SUB(k, P\varphi_i) = t_{i+1}$.
Then either $\tau(C\varphi_{i+1}) = \tau(C\varphi_i\{t_{i+1}/x_{i+1}\}) = \tau(C\varphi_i)$ in which case we have

$$\max\{\tau(C\varphi_{i+1}), \tau(D\varphi_{i+1})\} = \max\{\tau(C\varphi_i), \tau(D\varphi_i)\} \leq \max\{\tau(C), \tau(D)\} \text{ by (IH);}$$

or $\tau(C\varphi_{i+1}) = \tau(x_{i+1}, M\varphi_i) + \tau(t_{i+1})$
(note that by (IH) - A(i)-1
$\tau_{max}(x_{i+1}, M\varphi_i) = \tau_{min}(x_{i+1}, M\varphi_i) = \tau(x_{i+1}, M\varphi_i)$).

But by $SUB(k, P\varphi_i) = t_{i+1}$ it follows
$\tau_{min}(x_{i+1}, M\varphi_i) + \tau(t_{i+1}) \leq \tau(P\varphi_i)$ and in this case
$\tau(x_{i+1}, M\varphi_i) + \tau(t_{i+1}) \leq \tau(P\varphi_i) = \tau(P\varphi_{i+1})$.

Combining these inequalities with (IH) we get

$$\tau(C\varphi_{i+1}) \leq \max\{\tau(C), \tau(D)\}, \tau(D\varphi_{i+1}) = \tau(D\varphi_i) \leq \max\{\tau(C), \tau(D)\}.$$

Again the case $SUB(k, M\varphi_i) = t_{i+1}$, $SUB(k, D\varphi_i) = x_{i+1}$ is completely symmetric.
Thus A(i+1) is shown.

Q.E.D.

<u>THEOREM</u> 3. 4. 1.: Semantic clash resolution with negative setting is a decision procedure for VED.

<u>Proof</u>:
Let $\mathcal{C}$ be an arbitrary set of clauses.
We define $R^0(\mathcal{C}) = \mathcal{C}$

$$R^{i+1}(\mathcal{C}) = R^i(\mathcal{C}) \cup RSC\mathfrak{m}_n(R^i(\mathcal{C})).$$

If $\mathcal{C}$ is a set of Horn clauses then $R^*(\mathcal{C}) - \mathcal{C}$ consists of unit clauses only. In order to show finiteness of $R^*(\mathcal{C})/\sim_v$ it is sufficient to show that for all $C \in R^*(\mathcal{C})$ $\tau(C) \leq \tau(\mathcal{C})$.

We prove by induction on i that $\tau(R^i(\mathcal{C})) \leq \tau(\mathcal{C})$ and $R^i(\mathcal{C}) \in$ VED.

The case $i = 0$ is trivial.

(IH) Suppose that $\tau(R^i(\mathcal{C})) \leq \tau(\mathcal{C})$.

If $C \in R^{i+1}(\mathcal{C}) - R^i(\mathcal{C})$ then C is resolvent of a clash $(E; D_1, \dots, D_n)$ with $E, D_1, \dots, D_n \in R^i(\mathcal{C})$.

By an easy induction on i and by applying lemma 3.4.1 we get $\{E_{i+1}\} \in$ VED and $\tau(E_{i+1}) \leq \tau(\mathcal{C})$ from $\{E, D_{i+1}\} \in$ VED and $\tau(E_i) \leq \tau(\mathcal{C})$

But then $E_n = C$ and thus $\tau(C) \leq \tau(\mathcal{C})$ and $\{C\} \in$ VED.

It follows $\tau(R^{i+1}(\mathcal{C})) \leq \tau(\mathcal{C})$ and $R^{i+1}(\mathcal{C}) \in$ VED.

Therefore $R^*(\mathcal{C}) \in$ VED and $R^*(\mathcal{C})|\sim_v$ is finite.

Q.E.D.

In none of the decision problems studied in section 3.2 we made use of ordering structures for clauses. However, it is well known that in Horn logic fixed ordering of clauses is allowed, making clauses to lists rather than sets. If positive hyperresolution (**RSCm$_n$**) is combined with fixed ordering of rules and goals we come to HOSC - resolution (Horn ordered semantic clash resolution). HOSC - resolution was used in [LG 90] for problems of Horn clause implication, in [Lei90] also for deciding the classes KI, KII mentioned in chapter 3.2. Because set notation for clauses is not very practical in dealing with HOSC - resolution we create some special terminology.

Let $\{P, \neg P_1, \dots, \neg P_n\}$, $\{\neg Q_1, \dots, \neg Q_n\}$ be Horn clauses (a rule and a goal) s.t. there is a total ordering $<_C$ for negative literals in every clause C.

If $C = \{P, \neg P_1, \dots, \neg P_n\}$ and $\neg P_1 < \neg P_2 \ . \ . \ . < \neg P_n$ we write C as $P \leftarrow P_1, \dots, P_n$, what corresponds to the sequent notation usual in logic programming.

We define the HOSC - resolvent of a clash $(C; D_1, \dots, D_n)$ as follows:

$C = P \leftarrow P_1, \dots, P_n$ or $C = \leftarrow P_1, \dots, P_n$ and $D_i = Q_i \leftarrow$ for $i = 1, \dots, n$

(P_i, Q_i are atom formulas).

$R_0 = C$.

Suppose that $R_i = P' \leftarrow P_1', \dots, P_{n-i}'$ where P', P_k' are instances of P, P_k.
If $\{P_{n-i}', D_{i+1}\eta\}$ (for some renaming substitution η) is unifiable by m.g.u. σ

we set

$R_{i+1} = P'\sigma \leftarrow P_1'\sigma, \dots, P_{n-i-1}'\sigma$ for $i < n-1$

$R_{i+1} = P'\sigma \leftarrow$ for $i = n-1$

R_n is called HOSC-resolvent of the clash $(C; D_1, \dots, D_n)$ and is written as $HR(C; D_1, \dots, D_n)$.

Note that, if $C = \leftarrow P_1, \dots, P_n$ (a goal clause), then only $\Box$ can be $HR(C; D_1, \dots, D_n)$.

HOSC-resolution is a very efficient refinement because every clash-sequence can produce at most one resolvent; moreover it can be combined with forward subsumption. The set of all HOSC-resolvents definable by clash sequences from $\mathcal{C}$ is denoted by $HRS(\mathcal{C})$.

$HD(\mathcal{C})$ = is the set of all facts which can be derived by HOSC-resolution from $\mathcal{C}$.

More exactly: $H_0 = \text{Facts}(\mathcal{C})$

$H_{i+1} = H_i \cup HRS(\mathcal{C} \cup H_i)$ for all i and

$H^* = \bigcup_{i=0}^{\infty} H_i$, $HD(\mathcal{C}) = H^*|_{\sim_v}$

Note, that by the recursive definition also $\Box$ may be in $HD(\mathcal{C})$.

It was shown in [Lov78] that HOSC-resolution is complete. That means, for every set of Horn clauses $\mathcal{C}$:

$\mathcal{C}$ is unsatisfiable iff $\Box \in HD(\mathcal{C})$.

Note that $HD(\mathcal{C})$ is finite if all derivable facts have a common bound on term depth. If in the definition of the H_i only clauses of term depth $\le d$ are allowed, that is

$H_0' = \{C \mid C \in \text{Facts}(\mathcal{C}),\ \tau(C) \le d\}$

$H'_{i+1} = H_i' \cup \{C \mid C \in HRS(\mathcal{C} \cup H_i'),\ \tau(C) \le d\}$.

We get $H_d(\mathcal{C}) = \bigcup_{i=0}^{\infty} H_i'|_{\sim_v}$

as the set of facts derivable within depth d under equivalence w.r.t. renaming. $H_d(\mathcal{C})$ is always finite and can be constructed algorithmically. Note that $P\leftarrow\ \in H_d(\mathcal{C})$ is a decidable problem (P is an arbitrary atom), while

$P\leftarrow \in \{C \mid C \in HD(\mathcal{C}),\ \tau(C) \le d\}$ is not (otherwise, by $\tau(\Box) \le d$ the decision of $\Box \in \{C \mid C \in HD(\mathcal{C}),\ \tau(C) \le d\}$ would give a decision procedure for the satisfiability problem of Horn logic).

We now turn to the so called clause implication problem:

Let $\forall C$, $\forall D$ be the closed universal formulas corresponding to the clauses C and D. The problem, whether $\forall C \rightarrow \forall D$ is valid, is called the ***implication problem for clauses***. In [Sch88] it is proved that this problem is recursively unsolvable. Quite recently, Mycielski and Pacholski proved the unsolvability of the Horn clause implication problem [MP92] (i.e. the problem $\forall C \rightarrow \forall D$, where C is a Horn clause). Because of practical relevance to logic programming and in order to establish sharp borderlines between decidable and undecidable classes, an investigation of decidable subclasses of the Horn clause implication problem is of scientific interest. In [LG90], [Lei90], [Lei88] some decidable Horn classes for the implication problem have been investigated. The decision methods in [Lei88] were either ground methods or methods based on clause powers (iterated self-resolvents), while HOSC-resolution was used in [LG90], [Lei90]. Formally, the Horn clause implication problem is represented by a Horn set $\mathcal{C} = \{C, E_1, \dots, E_m\}$ where C is a Horn clause and the E_i are either ground facts or ground unit goals (the E_i stem from the Skolemization of $\neg\forall D$). The unsatisfiability of such a set of clauses is equivalent to the validity of the implication problem.

The case that C is a fact or a goal is trivial, because in both cases $\Box$ is the only possible clash resolvent. Thus the only interesting case is that C is a rule.

DEFINITION 3.4.2.: A Horn set $\mathcal{C}$ is called a ***non-trivial implication problem*** (NTI-Problem) if Facts($\mathcal{C}$) and Goals($\mathcal{C}$) consist of ground unit clauses only, $|\text{Rules}(\mathcal{C})| = 1$ and Facts($\mathcal{C}$) $\cup$ Goals($\mathcal{C}$) is consistent.

We now define the class which was called KIII in [Lei90]:

DEFINITION 3.4.3.: The class VH1 (one variable in the head) is the set of all NTI-problems $\mathcal{C}$ where the rule $C \in \mathcal{C}$ is subjected to the restrictions:

a) $|V(C_+)| \le 1$ and

b) there is a $P \in C_-$ s.t. $V(P) = V(C_-)$.

Remark: If $\mathcal{C}$ is in VH1 and for the rule $C \in \mathcal{C}$ we have $V(C_+) = 0$ then the satisfiability problem for $\mathcal{C}$ is trivial. In fact, if $C_+ = \{P\}$ then $HD(\mathcal{C}) \subseteq \{P \leftarrow, \square\} \cup Facts(\mathcal{C})$.

Hence, the only interesting case is $|V(C_+)| = 1$.

The class VH1, although it seems to be much more "special" than the class PVD, is in fact much harder to handle, the reason of which is that we do not get monotonic behaviour w.r.t. term depth.

Meanwhile the more general case for ≤ 2 variables in the head was solved by V. Rudenko [Rud 91]; his method is also based on HOSC-resolution.

There may be electrons with high term depth which are necessary for the derivation of H-resolvents with low term depth. Fortunately, we can give a bound on such possible "oscillations", and thus get decidability. The following lemma gives a characterization of clauses which are H-derivable from $\mathcal{C}$ for $\mathcal{C} \in$ VH1.

LEMMA 3.4.2.: Let $\mathcal{C}$ be a set of clauses in VH1 and let C be the rule in $\mathcal{C}$ with $C = P \leftarrow P_1, \dots, P_n$ s.t. $V(P_n) = V(C_-)$. Then either

a) $HD(\mathcal{C})$ is finite or

b) if $E' = HR(C; E_1, \dots, E_n)$ and $E_1 \in HD(\mathcal{C}) - Facts(\mathcal{C})$ s.t. $E_1 = P\lambda \leftarrow$ for $\lambda = \{s/x\}$ then $E' = P\mu \leftarrow$ where $\mu = \{t/x\}$ for a term t s. t. $\tau(s) < \tau(t)$.

Proof:

It is sufficient to focus on the case $|V(P)| = 1$, because for $|V(P)| = 0$ $HD(\mathcal{C})$ is trivially finite.

Thus suppose that $V(P) = \{x\}$. If $x \notin V(C_-)$ then $P\leftarrow$ is the only possible H-resolvent with nucleus C, and $HD(\mathcal{C})$ is finite.

So we may assume that $x \in V(C_-)$.

We proceed by analyzing $CORR(P,P_n)$:

Case 1) $CORR(P,P_n)$ contains a pair (s,x) for a term s.

Case 1.1) $CORR(P,P_n)$ contains a pair (s,x) s.t. s is a ground term.

Suppose that $Q \leftarrow \in HRS(\mathcal{C})$. Then, because C is the only rule in $\mathcal{C}$ and by $V(P) = \{x\}$, $Q = P\{t/x\}$ for some ground term t.

For abbreviation, we write $P[t]$ instead of $P\{t/x\}$.

Because s is ground, we have $(s,x) \in$ CORR(P[t],P_n). By V(P_n) = V($C\_$), every H-resolvent R← from a clash (C; $E_1, \ldots, E_n$) is determined by the first electron E_1 only; that means R = Pλ for λ = m.g.u. of {P_n←, E_1}. On the other hand, E_1 alone cannot guarantee the existence of a clash resolvent; here we have to take into account all of the E_i. Thus the resolvents are in HRS ($\mathcal{C}$) if the first electron is in Facts ($\mathcal{C}$). Therefore we may suppose that the first electron is of the form P[t]← with P[t]← ∈ HRS ($\mathcal{C}$).

So let E = HR(C; P[t]←, $E_2, \ldots, E_n$) be a clash resolvent with P[t]← ∈ HRS($\mathcal{C}$); this clash can only define a resolvent if {P [t],P_n} is unifiable.

By (s,x) ∈ CORR(P[t],P_n), every m.g.u. λ must fulfil sλ = xλ and by s ground, s/x ∈ λ. That means the clash resolvent E is P[s]←; because E = P[s]← for every choice of t, HD($\mathcal{C}$) must be finite.

Case 1.2) For all (s,x) ∈ CORR(P,P_n), s is not ground.
By V(P) = {x} we conclude that s contains x.
If s = x we get (x,x) ∈ CORR(P,P_n). Let P[t]← ∈ HRS ($\mathcal{C}$) and E be the clash resolvent of (C; P[t]←, $E_2, \ldots, E_n$); then { P [t], P_n } must be unifiable by a m.g.u. λ.
By (t,x) ∈ CORR(P[t], P_n) we conclude t/x ∈ λ (note that by x ∈ V($C\_$) t is ground). It follows that E = P[t] ← where P[t] ← ∈ HRS($\mathcal{C}$) and HD($\mathcal{C}$) is finite.
If s ≠ x then s = g[x] for a term g[x] properly containing x. Again, let E = HR(C; P[t] ←,$E_2, \ldots, E_n$) and λ be m.g.u. of {P[t],P_n}. By (g[x],x) ∈ CORR(P,P_n) we have
(g[t],x) ∈ CORR(P[t],P_n).
Because g[t] is ground we must have g[t]/x ∈ λ and E = P[g[t]].
By definition of the term g[x] we have $\tau(t) < \tau($ g[t]) and E fulfils condition b).

Case 2) CORR(P,P_n) ≠ {(P, P_n)}, but there is no pair s.t. (s,x) ∈ CORR(P,P_n).
Because P_n contains x there must be a term g [x] properly containing x and a term s s.t. (s, g[x]) ∈ CORR(P,P_n) and (s, g[x]) is irreducible (in the sense of [MM 82]). If s is not a variable then for every renaming substitution η, (sη, g[x]) is not unifiable. It follows that

for any ground instance P[t] $\{P[t], P_n\}$ is not unifiable. So no $E \in HRS(\mathcal{C})$ can be used as first electron in a clash, and $HD(\mathcal{C})$ must be finite.

If $s \in V$ then $s = x$ (by $V(P) = \{x\}$) and $(x, g[x]) \in CORR(P, P_n)$.

Let $g[x] = f(t_1, \dots, t_k)$ for an $f \in FS(\mathcal{C})$ and terms $t_1, \dots, t_k$.

If $E = HR(C;\ P[t] \leftarrow, E_2, \dots, E_n)$ then there is a m.g.u. λ of $\{P[t], P_n\}$.

By $(t, g[x]) \in CORR(P[t], P_n)$ and $g[x] = f(t_1, \dots, t_k)$ we conclude that $t = f(s_1, \dots, s_k)$ for some ground terms s_i (by $x \in V(P_n)$ and $P[t] \leftarrow\ \in HRS(\mathcal{C})$, t must be ground).

Because x must occur in some t_i, λ must contain r/x for a ground term r which is a subterm of an s_j.

It follows that $E = P[r]$ and $\tau(r) < \tau(t)$ and $\tau(E) \le \tau(P[t])$.

Let $d = \max\{\tau(E) \mid E \in HRS(\mathcal{C}) \cup Facts(\mathcal{C})\}$. Then $HD(\mathcal{C}) = HD_d(\mathcal{C})$; but $HD_d(\mathcal{C})$ is finite.

Case 3) $CORR(P, P_n) = \{(P, P_n)\}$.

By $V(P) \subseteq V(P_n)$ the first electron in a clash determines the form of the clash resolvent. But $CORR(P, P_n) = \{(P, P_n)\}$ implies that $\{P, P_n\}$ is not unifiable and no $P[t] \leftarrow\ \in HD(\mathcal{C}) - Facts(\mathcal{C})$ can be used as first electron. It follows that $|HD(\mathcal{C})| \le 2\,|Facts(\mathcal{C})| + 1$.

Combining the parts of the proof we get:
In cases 1.1), 2), 3) $HD(\mathcal{C})$ is finite, while property b) holds in case 1.2).

Q.E.D.

By the proof of lemma 4.1 we also get a method to decide whether a) or b) holds; for such a decision we simply have to analyze $CORR(P, P_n)$. In case a) we compute $HD(\mathcal{C})$. In case b) the situation is more difficult because $HD(\mathcal{C})$ may be infinite. But we will show that there is a bound on term depth d s.t. $\square \in HD(\mathcal{C})$ iff $\square \in HD_d(\mathcal{C})$.

<u>DEFINITION</u> 3.4.4.: Let $\mathcal{C}$ be a set of clauses in VH1 and let C be the rule in $\mathcal{C}$ with $C = P \leftarrow P_1, \dots, P_n$ and $V(P_n) = V(C\_)$.
If $\mathcal{C}$ fulfils condition b) in lemma 4.1 we say that $\mathcal{C}$ is *term depth increasing* (TDI).

<u>LEMMA</u> 3.4.3.: Let $\mathcal{C}$ be TDI and in VH1. Then there is a number d (depending recusively on $\mathcal{C}$) s.t. $\square \in HD(\mathcal{C})$ iff $\square \in HD_d(\mathcal{C})$.

<u>Proof:</u>

Let $C = P \leftarrow P_1, \dots, P_n$ be the single rule in $\mathcal{C}$ and $V(P_n) = V(C\_)$. We exclude the trivial cases $V(P) = 0$ or $V(P) \cap V(C\_) = \emptyset$.

Let $E = HR(C, (E_1, \dots, E_n))$ and $E_1 = Q \leftarrow$. Then E is determined by E_1 only (see also case 1.1 in lemma 4.1).

Obviously, the unifiability of $\{P_n, Q\}$ is a necessary condition for the existence of a resolvent of $(C; E_1, \dots, E_n)$. Thus, if $\{Q, P_n\}$ is unifiable by m.g.u. σ, we call $P\sigma \leftarrow$ a <u>*potential*</u> <u>*H-resolvent*</u> with first electron E_1. The term "potential" is justified because - in case of existence - the resolvent indeed is $P\sigma \leftarrow$.

Let $\Lambda = \{Q \leftarrow \mid Q \leftarrow \in \mathcal{C}$ and $\{Q, P_n\}$ is unifiable$\}$.

Λ is of crucial importance to $HD(\mathcal{C})$ because only facts from Λ, used as first electrons, can lead to successful derivations.

Let $\Lambda = \{Q_1 \leftarrow, \dots, Q_k \leftarrow\}$; then every $PE1(Q_i)$ (potential H-resolvent with first electron Q_i) is of the form $P[s_i] \leftarrow$ where $P[s_i] = P\{s_i/x\}$ and s_i is ground.

If $HRS(\mathcal{C}) = \emptyset$ then $HD(\mathcal{C}) = Facts(\mathcal{C})$ and $HD(\mathcal{C}) = HD_d(\mathcal{C})$ for $d = \tau(\mathcal{C})$.

If $\square \in HRS(\mathcal{C})$ there must exist a fact $Q \leftarrow \in Facts(\mathcal{C})$ and a goal $\leftarrow Q \in goals(\mathcal{C})$, and again we set $d = \tau(\mathcal{C})$ (note that a clash with nucleus C cannot give $\square$).

If $HRS(\mathcal{C}) \neq \emptyset$ and $\square \notin HRS(\mathcal{C})$ we proceed in the following way:

Let $A = \{i \mid P[s_i] \leftarrow \in HRS(\mathcal{C})\}$ and $B = \{1, \dots, k\} - A$.

a) $C = P \leftarrow P_1$.

In this case $B = \emptyset$, because P_1 also determines the existence (and not only the form) of the resolvent; so every $PE1(Q_i)$ is an actual resolvent.

Because $\square \notin HRS(\mathcal{C})$, $\square$ can only be in $HD(\mathcal{C})$ if there is a $P[s] \leftarrow \in HD(\mathcal{C})$ s.t. $\leftarrow P[s] \in Goals(\mathcal{C})$ (Note that all facts in $HD(\mathcal{C})$ must be ground by $V(P) \subseteq V(P_n)$).

Define $s_{i1} = s_i$ for $i = 1, \dots, k$

$s_{ik+1} = s_{ik}$ if $\{P[s_{ik}], P_n\}$ is not unifiable

$= t$ if $\{P[s_{ik}], P_n\}$ is unifiable by m.g.u. σ and $t = x\sigma$ ($P[t]$ is H-resolvent).

Because $\mathcal{C}$ is TDI we either have $s_{ik+1} = s_{ik}$ and no resolvent can be derived by (single) electron $P[s_{ik}] \leftarrow$ or $\tau(s_{ik+1}) > \tau(s_{ik})$.

If $\tau(P[s_{ij}]) > \tau(Goals(\mathcal{C}))$ then no clause derived from $P[s_{ij}] \leftarrow$ (using $P[s_{ij}] \leftarrow$ as electron in the derivation) can ever resolve with a clause in

Goals($\mathcal{C}$) because Goals($\mathcal{C}$) consists of ground clauses only. Thus we may choose $d = \tau($ Goals($\mathcal{C}$)).

b) $C = P \leftarrow P_1, \dots, P_n$ with $n \geq 2$.

Here we face a situation which is substantially more difficult:

1) $B \neq \emptyset$ is possible and

2) resolvents appearing as n-th electrons for $n > 1$ also determine the existence (but not the form) of further H-resolvents. Let

$\mathcal{C}_0$ = Facts($\mathcal{C}$),
$\mathcal{C}_{k+1}$ = HRS($\mathcal{C}_k \cup \mathcal{C}$) $\cup\, \mathcal{C}_k$ and
$\tau_1 = \max\{ \tau(Q_i) \mid i = 1, \dots, k \}$

By $V(P_n) = V(C\_)$ there is a number r s.t. for all substitutions ϑ $\tau(C\_\vartheta) \leq \tau(P_n\vartheta) + r$.

By the last property we conclude that a clash with first electron $Q_i \leftarrow$ can only be generated if all electrons in the clash are of depth $\leq \tau_1 + r$ ($\tau(C\_\vartheta) \leq \tau_1 + r$ by $P_n\vartheta = Q_i$ for some ϑ); note that for every ϑ with $P_n\vartheta = Q_i$ C_ϑ is ground. Moreover, a contradiction can only be derived if $P' \leftarrow\ \in$ HD($\mathcal{C}$) resolves with $\leftarrow P'$ in Goals($\mathcal{C}$). For such a P' we have $\tau(P') \leq \tau($ Goals($\mathcal{C}$)).

Again, let us define s_{ij}:

If $P[s_i] \leftarrow\ \in$ HRS($\mathcal{C}$) (or $i \in A$) then $s_{i1} = s_i$.

Suppose now that $P[s_{ij}] \leftarrow\ \in \mathcal{C}_j$ and there are $E_2, \dots, E_n \in \mathcal{C}_j$ and a clause E s.t. $E = HR(C; P[s_{ij}] \leftarrow, E_2, \dots, E_n)$ (by definition of the $\mathcal{C}_j$, E is in $\mathcal{C}_{j+1}$).

E must be of the form $P[t]$ for some ground term t. So in the case that E exists we set $s_{ij+1} = t$.

If there is no H-resolvent definable by first electron $P[s_{ij}] \leftarrow$ and facts in $\mathcal{C}_j$ then we set $s_{ij+1} = s_{ij}$.

Define $l_{ij} = \tau(s_{ij})$ if $P[s_{ij}] \leftarrow\ \in \mathcal{C}_j$
$\quad\quad = 0$ if $P[s_{ij}] \leftarrow\ \notin \mathcal{C}_j$ (or $P[s_i] \leftarrow\ \notin \mathcal{C}_j$).

We may represent the "status" of $\mathcal{C}_j$ by the tuple $(l_{1j}, \dots, l_{kj})$.

Suppose that for some $\mathcal{C}_j$ we have 1) $\square \notin \mathcal{C}_j$ and 2) for all $i = 1,\dots,k$ $l_{ij} = 0$ or $l_{ij} > \tau_2 = \max\{ \tau_1 + r, \tau(\text{goals}(\mathcal{C})) \}$.

Then $\square \notin \mathcal{C}_k$ for all $k \geq j$ and therefore $\square \notin HD(\mathcal{C})$. This holds because $l_{ik} \geq l_{ij}$ (note that $\mathcal{C}$ is TDI) for all $k \geq j$; so, for $k \geq j$, $P[s_{ik}] \leftarrow$ neither resolves with a goal in $\mathcal{C}$ nor can it contribute to a resolvable clash with first electron Q_i for $i \in B$ (where $l_{ij} = 0$); recall that C_σ is ground for any σ with $dom(\sigma) = V(P_n)$.

If there is some i s.t. $l_{ij} > 0$ and $l_{ij} \leq \tau_2$ (for a $\mathcal{C}_j$) then $P[s_{ij}] \leftarrow$ (potentially) contributes to new H-resolvents as n-th electron for $n > 1$. If $P[s_{ij}] \leftarrow$ is first electron in a clash, there may be an n-th electron ($n > 1$) of the form $P[s_{kj}] \leftarrow$ s.t. $l_{kj} > \tau_2$, and the clash is resolvable. On the other hand, every clause $P[s] \leftarrow$ s.t. $\tau(P[s]) > \tau(P[s_{ij}]) + r$ is useless in a clash with first electron $P[s_{ij}] \leftarrow$.

No $P[s_{ij}] \leftarrow$ with $l_{ij} \geq \tau_2 + k(r+1)$ can ever contribute to a derivation of $\square$; this can be explained in the following way:

Let $P[s_{ij}] \leftarrow \in \mathcal{C}_j$ with $l_{ij} \geq \tau_2 + k(r+1)$ and $(m_{1j},\dots,m_{kj})$ be the tupel $(l_{1j},\dots,l_{kj})$ after ordering the components of the last tupel under $\leq$. Then either $m_{ij} > \tau_2$ for all i or there is a maximal number p s.t. $m_{(pj)+1} - m_{pj} > r$; in the first case all relevant clauses are already exhausted and $\square \in \mathcal{C}_j$ iff there is a $k \geq j$ s.t. $\square \in \mathcal{C}_k$.

If $m_{(pj)+1} - m_{pj} > r$ s.t. p is maximal w.r.t. this property then every $P[s_{ij}] \leftarrow$ s.t. $l_{ij} \geq m_{(pj)+1}$ is useless as electron in a derivation of a $P[s] \leftarrow$ s.t. $\tau(s) \leq \tau_2$.

Thus it is sufficient to deal with clauses of depth $< \tau_2 + k(r+1)$ only, and by defining $d = \tau_2 + k(r+1)$ we get $\square \in HD(\mathcal{C})$ iff $\square \in HD_d(\mathcal{C})$.

Q.E.D.

<u>THEOREM</u> 3. 4. 2.: The satisfiability problem is decidable for the class VH1.

<u>Proof:</u>

Because H-resolution is complete, $\mathcal{C}$ is unsatisfiable iff $\square \in HD(\mathcal{C})$.

Suppose that $\mathcal{C}$ is in VH1 and that C is the rule in $\mathcal{C}$. If the literal $P' \in C\_$ with $V(P) = V(C\_)$ is not the last literal in $C\_$, we may reorder C to $P \leftarrow P_1,\dots,P_n$

s.t. $P_n = P'$ without loosing completeness. So we may assume that $C = P \leftarrow P_1,\dots,P_n$ and $V(P_n) = V(C_-)$. By the lemmas 3.4.2, and 3.4.3 we conclude that either $HD(\mathcal{C})$ is finite or there is a number d (which can be computed effectively from $\mathcal{C}$) s.t. $\square \in HD(\mathcal{C})$ iff $\square \in HD_d(\mathcal{C})$. Because it is decidable which of both cases applies, we either compute $HD(\mathcal{C})$ or $HD_d(\mathcal{C})$, both being finite. The resulting decision procedure is obvious.

Q.E.D.

<u>Corollary</u> 3.4.1.: Let VH1B2 (B2 for two variables in the body) be the class of all non-trivial implication problems s.t. the rule C fulfils $|V(C_+)| \leq 1$ and $|V(C_-)| \leq 2$. Then VH1B2 is decidable.

<u>Proof:</u>

Let $\mathcal{C} \in$ VH1B2 and C be the rule clause in $\mathcal{C}$.

case a: $|V(C_-)| \leq 1$. Then $V(C_-) = V(P)$ for some $P \in C_-$ and $\mathcal{C} \in$ VH1.

case b: $|V(C_-)| = 2$.

case b_1: There is a $P \in C_-$ s.t. $|V(P)| = 2$. Then $V(P) = V(C_-)$ and - again - $\mathcal{C} \in$ VH1.

case b_2: For all $P \in C_-$ it holds $|V(P)| \leq 1$. Suppose that $V(C_-) = \{x,y\}$,

$K_0 = \{P_1,\dots,P_k\}$ = set of ground atoms in C_- ,
$K_1 = \{Q_1,\dots,Q_l\}$ = set of atoms in C_- containing x,
$K_2 = \{R_1,\dots,R_m\}$ set of atoms in C_- containing y.
Note that $K_1 \cap K_2 = \emptyset$.

Because $|V(C_+)| \leq 1$ we get $V(C_+) \cap V(K_1) = \emptyset$ or $V(C_+) \cap V(K_2) = \emptyset$.

Let w.l.o.g. $V(C_+) \cap V(K_2) = \emptyset$.
Then $V(C_+ \cup K_0 \cup K_1) \cap V(K_2) = \emptyset$ and C can be decomposed.
More exactly, let $C_1 = P \leftarrow P_1,\dots,P_k,Q_1,\dots,Q_l$ (for $C_+ = \{P\}$), $C_2 = \leftarrow R_1,\dots,R_m$ and

$$\mathcal{C}_1 = (\mathcal{C} - \{C\}) \cup \{C_1\},\quad \mathcal{C}_2 = (\mathcal{C} - \{C\}) \cup \{C_2\}.$$

Obviously, $\mathcal{C}$ is unsatisfiable iff both $\mathcal{C}_1$ and $\mathcal{C}_2$ are unsatisfiable. But $\mathcal{C}_1$ is in VH1 and $\mathcal{C}_2$ is a trivial implication problem.

Q.E.D.

For deciding VH1 we didn't use full HOSC-resolution, but a modification which cuts terms at some depth d, which is computed from the structure of the implication problem. While this modification always terminates, it is not complete. The problem was that only variable occurrences, but not the term structure itself was restricted in VH1. To compute $HD_d(\mathcal{C})$ for some depth d is much more a "classical" decision algorithm than a theorem prover. The method was refined by V. Rudenko [Rud91] by defining other complexity measures for atoms than term depth; by this method he succeeded to show, for the case $|V(C_+)| \leq 2$, $V(C) = V(L)$ for a $L \in C_-$, that either $HD(\mathcal{C})$ is finite or $HD(\mathcal{C})$ defines a monotonically increasing chain (w.r.t complexity) of facts; thus he obtained an improved decision algorithm for deciding Horn implication problems. His method shows that HOSC-resolution is a good basis structure to design decision algorithms for such problems. It seems to be that the problem gets substantially more difficult if only $|V(C_+)| \leq 2$ is stated in the definition, without restricting variable occurrence in C_-. The more general case, $V(L) = V(C)$ for an $L \in C_-$, without restriction on the number of variables in the head, is still unsolved.

CHAPTER 4

COMPLETENESS OF ORDERING REFINEMENTS

As soon as the resolution method was proposed in [Rob 65] and the analogous 'inverse method' in [Mas 64], it was noticed that for the completeness of these methods it is not always required that all the literals in a clause are resolved upon. Prohibiting resolution on some literals in a clause immediately means a reduction of the branching factor for the search tree. This reduction often leads to a slower growth of the set of derived clauses, which in turn can lead to finding the empty clause faster. For some classes of first order predicate logic formulas, it is possible to find such refinements of resolution that the search space turns out to be finite, thus giving a decision algorithm for these classes.

DEFINITION 4.1.: A resolution refinement is called complete iff the following holds: if a set of clauses is unsatisfiable, an empty clause is derivable from the set by the resolution refinement.

Quite obviously, we are mainly interested in complete refinements. However, sometimes the refinement we are looking for is generally incomplete, hence does not guarantee finding an empty clause for all unsatisfiable clause sets, yet is complete for a certain class of sets of clauses . In this case we can use the refinement, provided that the set of clauses we are investigating belongs to this class.

During the current chapter a clause is assumed to be a list rather than a set; that is, a clause can contain two syntactically equal literals. Consider some clause $C = (L_1, L_2, \dots, L_n)$ and such a restriction or refinement of binary resolution that the literals $L_1, L_2, \dots, L_k$ in C are allowed to be resolved upon and the literals $L_{k+1}, \dots, L_n$ are not allowed to be resolved upon. We can say that this refinement is defined by a two-place predicate '>' defined on literals in the same clause with the meaning that whenever $(L_i > L_j)$ is true for some literals L_i and L_j in the clause, L_i is preferred over L_j, that is, L_j is not allowed to be resolved upon. The question whether L_i is allowed to be resolved upon, depends on whether the given clause contains any literal L_q such that ($L_q > L_i$) holds: if this is the case, then L_i cannot be resolved upon in this clause, otherwise it can be resolved upon.

EXAMPLE 4.1.: In clause C for any two literals L_i, L_j the predicate ($L_i > L_j$) holds iff $i \leq k$ And $j > k$.

In order that the restriction were completely defined for a class of formulas, the predicate '>' must be defined so that for any clause C derivable from any set of clauses S in the given class by the restriction defined by '>', '>' were computable for any pair of literals in C. Generally '>' might not depend on the syntactic form of literals only; the derivation tree of the literal could also be taken into account (lock-resolution is a prime example of taking the derivation into account).

4.1 PROVING COMPLETENESS BY LIFTING

One of the usual methods for proving the completeness of any variant of the resolution method consists of the following steps (see [CL 73]):
(From now on we will consider completeness proofs of such kind and only for binary resolution, where the resolution and factoring steps are performed separately).

Ground case: It is proved that the variant is complete for the ground case, that is, for the Herbrand expansion of the set of clauses. Instead of using semantic trees we will use the completeness proof for ground lock-resolution (see [Boy 71], [CL 73]), where an induction on the k-parameter is used (k-parameter is the difference between the sum of lengths of clauses and the number of clauses).

Lifting: Let A and B be two non-ground clauses: $A = (A_1, ..., A_n)$, $B = (B_1, ..., B_m)$. Let A' and B' be any corresponding ground clauses from the Herbrand expansion of A and B. Lifting the ground completeness proof to the variable-containing case means showing that whenever a new clause C' can be inferred from A' and B' by some complete refinement of resolution, such a clause C can be inferred from A and B so that C' is a ground instance of C.

Thus clauses with variables can be seen as a 'shorthand' for a possibly infinite set of ground clauses with the same structure. A resolution step for non-ground clauses can be seen as a shorthand for a possibly infinite number of resolution steps for subsets of corresponding ground clauses.

In chapter 5 C.Fermüller investigates resolution variants, the completeness of which he proves by using the method of semantic trees introduced in [KH 69]; in several cases the last method turns out to have some special properties, simplifying the completeness proof.

It is clear that for a completeness proof for any class of formulas it suffices to take into account only those ground clauses which can be obtained from the clausal form of some formula in the class. E.g., for the Monadic Class it suffices to consider clauses consisting of one-place literals only.

4.2 SOME IMPORTANT PROPERTIES OF THE '>'-PREDICATES

4.2.1 Completeness for the ground case

When proving the completeness of any refinement of resolution given by a certain '>'-predicate, the first step is to prove completeness for the ground case (remember that in this chapter we consider only proofs with the structure presented in section 4.1).

Not all possible '>'-predicates yield a complete refinement for the ground case. Obviously, the first thing to note is that generally it is needed that in every clause there is at least one literal which is allowed to be resolved upon:

(A) for each clause $C = (L_1, \ldots, L_n)$ there is at least one i such that $L_j \not> L_i$ holds for each j, $1 \leq j \leq n$.

Notice that since for deriving an empty clause from any unsatisfiable set of clauses we have to use clauses with a single literal (assuming we treat factorization as a separate rule), the fact (L $\not>$ L) for all literals L follows from condition (A).

As demonstrated by the following example, the property (A) does not suffice for the general completeness of a '>'-refinement for ground clauses.

EXAMPLE 4.2.: (taken from [Boy 71]). Consider the following unsatisfiable set of ground clauses:

1: (P, R)
2: (¬R, P)
3: (R, ¬P)
4: (¬P, ¬R)

Let us allow to resolve only upon the first literal of each clause. On the first level we can thus get only the following two clauses, which both are tautologies:

1, 4 give 5: (R, ¬R)
2, 3 give 6: (P, ¬P)

We allow all literals of these tautologies to be resolved upon. Thus we get:

6, 1 give 7: (P, R)
5, 2 give 8: (¬R, P)
5, 3 give 9: (R, ¬P)
6, 4 give 10: (¬P, ¬R)

The set we got on the second level is identical to the input set. If we again allow to resolve only upon the first literals, no new clauses can be derived, thus the empty clause cannot be derived.

The problem with the '>'-refinement used in the example is that the '>' predicate is cyclical. Consider the definition of the '>'-predicate we used in terms of the syntax of literals:

(P > R) and (R > ¬P) and (¬P > ¬R) and (¬R > P)

The cycle starts with 'P' and ends with 'P'. Since the '>'- predicate defined in the above way is not transitive (e.g, although $(P > R)$ and $(R > \neg P)$ holds, $(P > \neg P)$ does not hold), condition (A) holds for clauses which do not contain all the literals P, R, $\neg P$, $\neg R$.

If the given '>'-predicate is transitive, that is, the following condition holds:

(B) for all literals X, Y, Z: $(X > Y)$ and $(Y > Z)$ implies $(X > Z)$

then for every literal X in the cycle the fact $(X > X)$ must hold. In this case the condition (A) cannot be fulfilled. Thus, if we have a '>'-predicate for which both condition (A) and transitivity (condition (B)) hold, then '>' cannot form cycles.

The opposite is not true, however: although acyclicity guarantees (A), it does not guarantee (B). For example, the >'-predicate defined by $((P > R)$ and $(R > \neg R))$ is acyclical, but not transitive, as $(P > \neg R)$ does not hold.

Nevertheless, every acyclical '>'-predicate can be strenghtened to a new '>'-predicate for which both, (A) and (B), hold. By strengthening the '>'-predicate we mean forming the transitive closure of the '>'-predicate, that is, adding new pairs of literals for which '>'-holds, while retaining all the old pairs. E.g., the '>'-predicate defined by $((P > R)$ And $(R > \neg R))$ is strengthened to a transitive predicate by adding $(P > \neg R)$.

In the following we will only investigate such '>'- predicates, for which both conditions, (A) and (B), hold for the ground case. We will call such '>'-predicates orderings.

As shown by S. Maslov (see [Mas 68], [Mas 71]), the acyclicity of '>' (therefore also conditions (A) and (B)) is enough for the resolution with a '>'-refinement to be complete for ground formulas.

Let us consider some orderings. We will say that the ordering is complete iff the following condition holds:

(C) for all ground literals X and Y with $X \neq Y$ either $(X > Y)$ or $(Y > X)$ holds.

<u>EXAMPLE</u> 4.3.: $(P > R > \neg P > \neg R)$ is a complete ordering for a clause set containing only atoms R and P.

As an example of an incomplete ordering we could define:

$$P > \neg P, \quad P > \neg R, \quad R > \neg P, \quad R > \neg R.$$

Then none of

$$P > R, \quad R > P, \quad \neg P > \neg R, \quad \neg R > \neg P$$

holds and in the clauses (P, R) and ($\neg$P, $\neg$ R) both literals are allowed to be resolved upon.

We will write (X >< Y) to denote the incomparability of any literals X and Y. By the definition of orderings (X >< X) holds for any literal X and any ordering. In the last example of '>' we have (P >< R) and ($\neg$P >< $\neg$R).

An important class of orderings is formed by lock-resolution (introduced in [Boy 71]). These orderings are not defined in terms of syntax of literals, but in terms of a derivation tree. Namely, for each literal in the input set of clauses a certain numerical index is given and the ordering is defined in terms of these indices:

$$X > Y \quad \text{iff} \quad \text{index}(X) < \text{index}(Y)$$

We will denote the fact that the literal A has index i by writing A{i}. Literals in the derived clauses retain the same index they had in the parent clauses. We will not consider an index to be a part of the literal; instead we consider it to be a part of a deduction tree.

Thus syntactically equal literals in different clauses are treated as different objects from the standpoint of lock resolution. The fact that a lock-ordering is indeed an ordering (that is, enjoys properties (A) and (B)) follows from the fact that the usual ordering '<' of natural numbers has properties (A) and (B).

Since we know that lock-resolution is complete, it is simple to prove that resolution for ground clauses with any syntactically defined acyclical '>-'predicate ordering is complete. We start by strengthening the acyclical '>'-predicate to the corresponding ordering by adding new pairs of literals for which '>' holds, as shown earlier. The restriction given by this ordering is stronger than the one given by the original '>'- predicate, so whenever the ordering gives a complete resolution refinement, the original '>'-predicate also does. The ordering we get may contain several trees and branches of comparable literals, literals in different chains and different branches being incomparable. When we here speak of "trees", we mean oriented graphs which do not contain oriented cycles, although they may contain unoriented cycles.

EXAMPLE 4. 4.:

(4. 1) ((P > R) and (¬P > ¬R))

(4. 2) ((P > R) and (R > ¬ P) and (P > ¬ P))

(4. 3) ((P > R) and (R > ¬P) and (P > ¬P) and (R > ¬R) and (P > ¬R))

are such orderings, the first consisting of two incomparable trees, the second containing an unoriented cycle and the third containing two incomparable branches.

We are going to give numerical indexes to all literals, such that the corresponding lock-strategy is stronger (or of exactly the same strength) than the strategy given by ordering. The indexing method is the following:

- all syntactically equal literals get the same index.
- trees are indexed one-by-one, that is, all indexes in the next tree are bigger than the indexes in the previous tree.

EXAMPLE 4. 5.:

((P{1} > R{2}) and (¬P{3} > ¬R{4})).

- we divide all vertexes between levels, so that vertexes at the next level have the same index, which is bigger than the index for the previous level.

EXAMPLE 4. 6.: ((P{1} > R{2}) and (R{2} > ¬P{3}) and (P{1} > ¬P{3})) and
((P{1} > R{2}) and (R{2} > ¬P{3}) and
(P{1} > ¬P{3}) and (R{2} > ¬R{4}) and (P{1} > ¬R{4}))

Such an indexing is possible for all orderings (recall properties (A) and (B)) and for all finite sets of literals. So, the fact that any acyclical '>'-ordering gives a complete resolution refinement for finite sets of clauses is proved. Notice that we never use infinite sets of clauses. Although the Herbrand expansion of a non-ground formula may be infinite, whenever this formula is unsatisfiable, the result can be shown using the first N clauses in the Herbrand expansion for some N. Generally we do not know N beforehand, but we know that iff the formula is unsatisfiable, the required finite Herbrand expansion can be formed.

An important point to notice is the difference between orderings defined in terms of syntax and in terms of deduction tree. This becomes noticeable when we investigate the treatment of tautological clauses and subsumed clauses.

<u>DEFINITION</u> 4. 2.: A clause is called tautological iff it contains some literal X and its negation X^d. A ground clause C subsumes a ground clause D iff the set of literals in C is a subset of the set of literals in D.

During the process of deriving new clauses by resolution we want to avoid any unnecessary derivations. As it is the case that for most of practically interesting formulas most of the derived clauses are subsumed by older clauses or subsume some older clauses, it is very important that we could avoid using any subsumed clauses. We have to consider the following question: in which cases does the resolution with ordering strategy allow us to discard subsumed clauses ?

It is useful to consider 'back subsumption' and 'forward subsumption' separately. By 'back subsumption' we mean the following restriction: whenever a newly derived clause C subsumes some older clause D, we eliminate D (that is, from now on we will not use D for any resolution steps). By 'forward subsumption' we mean the following: whenever a newly derived clause C is subsumed by some older clause D, we eliminate C.

We also consider separately the cases where a subsuming clause is syntactically equal to the subsumed clause (up to renaming variables) and the case where the subsuming clause is more general (for ground case this means to be shorter) than the subsumed clause. The first kind of subsumption we call 'trivial subsumption' and the second kind 'proper subsumption'.

By the compatibility of some kind of subsumption with some ordering we will mean that the combination of the resolution refinement defined by the ordering with the clause deletion strategy (using the kind of subsumption) is a complete resolution variant.

THEOREM 4.1.: Proper back subsumption is compatible with all orderings, including lock resolution.

Proof:

Recall that we use resolution as a process of iteratively generating new (without subsumption, strictly bigger) conjunctions of clauses. Consider an unsatisfiable set of clauses $S = \{A_1, A_2, ..., A_n\}$. Then the set $S' = \{A'_1, A_2, ..., A_n\}$ such that A'_1 subsumes A_1, is also unsatisfiable. Suppose A'_1 is used to properly back subsume A_1. Then at the next level the obtained set remains unsatisfiable. As the subsumption application was of a proper kind, the chain defined by replacing clauses by subsuming ones cannot be infinite: at every such step we get a more general clause, up to a one-literal clause (where in the non-ground case all the arguments of the predicate are different variables).

Q.E.D.

For forward subsumption the proof does not work in the general case since even the proper forward subsumption can result in a cycle: although the set of clauses remains unsatisfiable, resolution may give a set of clauses which is identical to the set found at the last level.

However, both forward and backward subsumption are compatible with syntactically defined orderings (and acyclical '>'- predicates also) for the ground case, as shown in [Mas 68]. Again, this becomes immediately visible when we investigate subsumption for lock resolution.

As demonstrated in [Boy71], general subsumption is uncompatible with lock resolution even for the ground case:

1: (P{1}, R{2})

2: (¬R{3}, P{4})

3: (R{5}, ¬P{6})

4: (¬P{7}, ¬R{8})

At the first level we can thus only get the following two clauses, which both are tautologies:

1, 4 give 5: (R{2}, ¬R{8})

2, 3 give 6: (P{4}, ¬P{6})

Both clauses obtained at the next step are forward subsumed by older clauses:

5, 2 give 7: (¬R{8}, P{4})

6, 4 give 8: (¬P{6}, ¬R{8})

If we do not eliminate subsumed clauses resolution will proceed as follows:

7, 8 give

9: (¬R{8})

6, 8 give 10: (¬P{6}, ¬R{8})

9, 3 give 11: (¬P{6})

11, 1 give 12: (R{2})

9, 12 give □

We will examine how we can restrict forward subsumption so that it becomes compatible with lock resolution.

<u>THEOREM</u> 4. 2.: Forward subsumption with the following restriction is compatible with lock-resolution: A is defined to subsume B in the restricted sense iff A subsumes B in the ordinary sense and the corresponding literals in A and B have the same indices.

<u>Proof</u>:

Consider the deduction tree of an empty clause from some set of clauses by lock-resolution. Replace every occurrence of clause B by clause A in this tree. Consider two cases:

- A is syntactically equal to B (trivial subsumption). Then, since the indices are the same for A and B, the proof tree is unchanged and obviously all the steps remain permissible.
- A is more general than B (for the ground case this means that A is shorter than B). If A is shorter than B, then we have to cut off branches in the old tree corresponding to literals present in B, but not in A. For clauses in the tree for which B was an ancestor, we have also to cut off these literals

which were introduced by the branches we have cut off. Then, due to the restriction on the indices all the steps remain permissible lock-resolution steps. The essential possible change from the old tree is that the empty clause may be derived earlier in the rebuilt tree.

For non-ground clauses we note that replacing A by a more general clause permits all unifications in the rebuilt tree. Since indices do not depend on the syntax of the clause, the previous argument holds.

Q.E.D.

Consider some syntactically defined ordering for the ground case. As we know, this ordering can be modelled by lock resolution. Now, whenever some clause A subsumes a clause B, all the literals in A are also in B, thus their indices are equal (since the indices of syntactically equal literals are the same). Thus the restriction for indices in theorem 2 is always fulfilled and full subsumption is compatible with any syntactically defined ordering for the ground case.

In the light of subsumption the treatment of tautological clauses is trivial: we can always eliminate tautological clauses, provided that any clause derived with a tautology as one of the parents is forward subsumed by its second parent. For syntactically defined orderings for the ground case we can therefore eliminate all tautologies. This is not the case for lock-resolution, since, for example, (P{1}, R{2}) and (¬P{3}, P{3}) give (P{3}, R{2}), which has indices different from the parent clauses (see also the previous example).

4.2.2 Properties of orderings for the non-ground case

As a clause with variables can be considered to be a shorthand for a possibly infinite set of ground clauses (Herbrand expansion), the most important thing is to consider a relation among an ordering on a clause with variables and the ordering on all ground instances of this clause.

As said before, all the results of the previous chapter which hold for finite input sets of clauses, remain valid for the possibly infinite Herbrand expansion of any non-ground set of clauses.

For unrestricted binary resolution the lifting of a completeness proof to the non-ground level is possible due to the existence of a most general unifier for any two literals. For a resolution refinement it might happen that a resolution step which is possible for ground instances of clauses is not any more possible for the corresponding non-ground case, thus destroying lifting.

<u>EXAMPLE</u> 4.7.: Consider the following syntactically defined ordering for the general case:

($A >_{ig} B$) iff A is ground and B is non-ground.

Then only the first literals of the following clauses are allowed to be resolved upon:

(P(a), R(x)), (¬R(a), P(x)).

Therefore these two clauses cannot be used to derive any new clauses. However, for the ground case corresponding to these clauses the derivation of new clauses is possible:

(P(a), R(a)) and (¬R(a), P(a)) give (P(a), P(a)).

Indeed, if lifting is not preserved, we have no guarantee for the completeness of the corresponding ordering strategy. Consider the same ordering $>_{ig}$ and the following unsatisfiable set of clauses:

1: (P(a), R(x))
2: (¬R(a), P(x))
3: (R(a), ¬P(x))
4: (¬P(a), ¬R(x))

At the first level we can only get the following two clauses, which both are tautologies:

1, 4 give 5: (R(x), ¬R(x))
2, 3 give 6: (P(x), ¬P(x))

We allow all literals of these tautologies to be resolved upon. The set we get at the second level is identical to the input set, therefore the empty clause is never derived.

We want to separate a class of orderings which preserve lifting. Since conditions (A) and (B) hold for orderings, the following condition is what we need:

(D) for any literals X, Y and any ground substitution σ the following implication holds: $(X > Y) \Rightarrow (X\sigma > Y\sigma)$.

The following does not suffice:

(Dweak) $(X\sigma > Y\sigma) \Rightarrow \neg(Y > X)$.

The counterpositive of (Dweak) is $(Y > X) \Rightarrow \neg(X\sigma > Y\sigma)$, and this obviously holds for the ordering in our last example, since it follows from
$(Y >_{ig} X) \Rightarrow (X\sigma >\!\!<_{ig} Y\sigma)$. But in our example all the ground literals were incomparable w.r.t. the ordering $>_{ig}$.
Given any ordering obeying (D), it is possible to define the resolution refinement for this ordering in two distinct ways, so that lifting is preserved for both.

The first way is to order any clause with the given ordering and prohibit resolution on non-maximal literals, exactly as was done for the ground case. We will call such refinements "apriori refinements".

The second way, introduced in [Joy76], relies on noticing that in order to preserve lifting, it suffices to consider clauses under application of m.g.u. s. Consider two clauses,
$C_1 = (L_1, L_2, \ldots ,L_n)$ and
$C_2 = (G_1, G_2, \ldots , G_m)$, such that L_1 and G_1^d are unifiable with the substitution $\sigma = \text{m.g.u}(L_1, G_1^d)$. Given any ordering $>_x$ we can define a liftable refinement of binary resolution in the following way:
The clause $(L_2, \ldots , L_n, G_2, \ldots , G_m)\sigma$ is allowed to be derived by binary resolution iff there is no L_i in C_1 such that $(L_i\sigma >_x L_1\sigma)$, and no G_i in C_2 such that $(G_i\sigma >_x G_1\sigma)$.

The last kind of resolution refinements we will call "aposteriori refinements". As shown by C. Fermüller in [Fer91] and in chapter 5, in several cases aposteriori refinements can be much more powerful and convenient than apriori refinements. On the other hand, using aposteriori refinements in the actual proof search is connected with a noticeable computational overhead. Therefore, in case there exists an apriori refinement with the same restrictive power as the aposteriori refinement, it is better, because more efficient, to use the apriori refinement.

The fact that any syntactically defined acyclical '$>$'-predicate with property (D) gives us a generally complete resolution refinement was first proved by S.Maslov (see [Mas 68], [Mas 71]; the proof is presented in [Mas 64]). S. Maslov called such predicates "π- orderings". He also demonstrated that subsumption is compatible with π-orderings. As in the current chapter we have reserved the name 'ordering' for '$>$'-predicates with the properties (A) and (B) only, π-ordering is a wider class of predicates than the orderings in our sense. Yet we prefer not to drop the name 'π-orderings', so the reader is asked to keep

in mind that a 'π-ordering', differently from simply 'ordering', need not satisfy (B) (acyclicity suffices instead of (B)).

Due to (D) we can lift the completeness and subsumption-compatibility result for resolution with any syntactically defined acyclical '$>$'- predicate (see section 4.2.1). This gives the completeness and subsumption-compatibility proof for resolution with any π-ordering.

Notice that property (D) also holds for lock-resolution. Thus we get completeness for general (that is, not ground-only) lock-resolution by lifting. Theorems 1 and 2 were already formulated for the general case, so we know that these types of subsumption are compatible with general lock-resolution. An important point to notice is that, differently from the ground case, it is not possible to use lock-resolution to model any syntactically defined ordering with the property (D). We will give an example later.

Independently from S. Maslov, R. Kowalski and P. Hayes introduced orderings called A-orderings in the paper [KH 69]. An A-ordering is a '$>$'-predicate on atoms (not on literals!) which satisfies (A), (B) and (D). As A-orderings form a subclass of π-orderings, the completeness of a resolution refinement defined by an A-ordering follows from the completeness of resolution with π-orderings. We refer to chapter 5 for a more detailed investigation of A - ordering.

4.2.3 Factoring

In the ground case the notion of factoring is of no special importance, as the clauses are normally considered to be sets of ground literals and therefore the factoring operation is, so to say, performed implicitly by definition. This is not completely true for non- syntactically defined orderings, like lock-resolution, as syntactically equal ground literals can have different indices. For lock resolution it is therefore defined explicitly that in case a clause C contains two syntactically equal literals A_i and B_j with different indices i and j, C should be replaced by a clause where B_j is removed and the index i in A is replaced by the max(i, j).

For non-ground clauses the notion of factoring gains explicit importance, as set formation is not enough any more:

e.g, the clause (P(f(x)), P(y)) has a ground instance (P(f(a)), P(f(a))), which gives a one-element set P(f(a)). Obviously, in order to use lifting for completeness proofs, factoring has to be performed.

As noted in [Lei89] and in chapter 2, in theorem-proving literature factoring is treated in one of the two ways: Robinson's resolution method requires factoring during a binary resolution step only, and the set of factorized literals in a clause has to contain the literal which is resolved upon. The other way to treat factorization is to use it as a separate rule - this can be useful from the standpoint of implementation, as using factorization during a binary resolution step is connected with a certain computational overhead.

For ordering refinements of Robinson's resolution the set of factored literals in a clause always contains some literal L such that for no R in the clause $R >_0 L$. The same restriction can therefore be introduced to factoring as a separate rule: for any ordering $>_0$ it is required that any set of factored literals in a clause must contain a literal L such that for no other literal R in the same clause $R >_0 L$ holds.

4.3 POSSIBILITIES OF USING π-REFINEMENTS AS DECISION ALGORITHMS

By "π-refinement" we will mean an apriori resolution refinement based on some π-ordering.

We will present the following ordering to be used later:

DEFINITION 4.3.1: $(X >_{sub} Y)$ iff all arguments of the literal Y are proper subterms of some arguments of the literal X.

Then $P(f(y)) >_{sub} R(y,y)$, $R(g(f(y)),y) >_{sub} R(y, f(y))$,
$P(g(f(a)), a) >_{sub} R(a, f(a))$, $P(x, y) \not>_{sub} P(x)$, to give some few examples.

As shown by N. Zamov and V. Sharonov (e.g. in [ZS 74]), and independently by W. Joyner in [Joy 76], the apriori restriction strategy based on this ordering is a decision algorithm for clause forms of formulas from the Ackermann Class (prefix $\forall\exists\ldots\exists$) and Monadic Class (formulas with one-place predicates only). Notice that both of these classes are defined for the predicate logic without function symbols; the only function symbols appearing in clause forms of such formulas are the ones produced by Skolemization.

In this chapter we concentrate on completeness, so the reader is suggested to look at the following chapters to see the decision proofs. All such proofs show that the set of clauses derivable by the investigated strategy is finite for a clause form of any formula in these classes. This is achieved by showing that the term depth does not grow and due to condensing the length of derived clauses is also bounded (we call the factors subsuming their parents 'condensings') . So, if resolution does not produce any more clauses and it has not derived an empty clause, we can conclude (due to completeness of the strategy) that the input set of clauses must be satisfiable.

Let us investigate the properties of the $>_{sub}$-ordering. We can easily notice that this ordering relation has both, properties (A) and (B), thus is really an ordering. Therefore it is complete for the ground case. As $(P(a) >\!\!<_{sub} P(b))$, (C) does not hold. Consider property (D). Whenever $(X >_{sub} Y)$ for some literals X and Y, $(X\sigma >_{sub} Y\sigma)$ also holds. Thus we have proved that the given ordering is indeed a π-ordering.

We give an example of using the apriori restriction with a given ordering $>_{sub}$ as a decision strategy. Consider the following set of clauses:

1: (P(f(x)), P(x))

2: (¬P(f(x)), ¬P(x))

Ordering $>_{sub}$ allows to resolve only on the first literals, thus we can produce a single clause, (P(x), ¬P(x)), which is tautological and due to the fact that we used a π-refinement, we can eliminate it. As no more clauses can be derived, we know that the input set is satisfiable.

As one can easily check out, unrestricted binary resolution will deduce an infinite amount of clauses from the given set. As can be almost as easily checked, the same happens for lock- resolution with any indexing of the input clauses (recall that tautologies cannot be eliminated for lock-resolution). So, π- refinements cannot be modelled by lock-resolution.

4.4 BEYOND π-REFINEMENTS

As noticed by N. Zamov, V. Sharonov and W. Joyner, π- refinements (or, in Joyner's case, refinements based on A-orderings), cannot be used to decide the Gödel's Class, the class of formulas with the prefix $\forall\forall\exists$. The problem here is that the clause form contains clauses like (P(x, y), P(x, x)), which are left unordered by any π- ordering. The free use of such clauses produces clauses like (P(f(x,y), z), P(x, y)), which start producing terms with unbounded depth.

<u>EXAMPLE</u> 4.8.: From (P(x, x), P(x, y)) and (¬P(f(x,y), f(x,y)), ¬P(x, y)) we get (P(f(x,y), z), ¬P(x, y)). From the last one and from (¬P(x, f(x,y)), P(f(x,y), f(x,y))) we get: (¬P(x,y), P(f(f(x,y), y'), f(f(x,y), y'))). In this way the depth of terms is going to grow unbounded.

It is easily seen that no π-ordering can order the clause (P(x,y), P(x, x)) so that the first literal is preferred. If an ordering prefers P(x, y) to P(x, x), then it does not satisfy (D), since in order to satisfy (D) it must prefer the first literal of (P(a, a), P(a, a)), the ground instance of the last clause. Then it either is not syntactical or does not satisfy (A).

4.4.1 Amending lifting by an additional saturation rule

In the papers [Zam 72, 89, 89a] the problem is overcame by defining an ordering which does not generally satisfy (D), and by introducing a special saturation rule to resolution, so that for resolution augmented with this saturation rule, lifting is possible. We will define an ordering which essentially captures the idea of the orderings for the Minimal Gödel Class defined in [Zam 89a].

<u>DEFINITION</u> 4.4.1: ($X >_{sv} Y$) iff one of the following holds:

(1) each argument of the literal Y is a proper subterm of some argument of the literal X.

(2 neither X nor Y contain functional terms with arguments, the set of variables in Y is a proper subset of variables in X and the set of variables in Y is non-empty.

The proof that this ordering indeed gives a decision strategy for the Minimal Gödel's Class can be found in the paper [Zam 89a] and in chapter 6 (indeed, the orderings defined there are somewhat different, but for the Minimal Gödel Class they give exactly the same restrictions).
Obviously, the last ordering $>_{sv}$ is syntactically defined and satisfies conditions (A) and (B). Notice that the condition (1) in the '$>_{sv}$'-ordering is exactly the same as the '$>_{sub}$'-ordering given in the previous example.
The main problem, as we saw already, lies in condition (2), which destroys lifting. One way to overcome this, previously suggested by N. Zamov and V. Sharonov, consists of adding a new derivation rule, saturation, to resolution. Saturation allows to derive certain substitution instances of clauses. Understandably the idea is that saturation always allows to derive only a finite amount of substitution instances, and generally, the less the better. The following saturation rule suggested by N. Zamov makes resolution with the $>_{sv}$-ordering complete for Minimal Gödel's Class:

(SAT1) If resolving upon some literal L in some clause C is prohibited only due to the fact that C contains some literal M with ($M > L$) by condition (2), then derive all substitution instances $C\sigma$ of this clause such that $\sigma = \{y/x\}$, for $x \in V(M) - V(L)$, $y \in V(L)$.

<u>EXAMPLE</u> 4.9.: From (P(x,y), R(y,y)) (SAT1) derives (P(y,y), R(y,y)). From (A(f(x, y)), P(x,y), R(y,y)) (SAT1) derives nothing, since resolution upon R(y,y) is prohibited due to condition (1) already.

Let us see why our last ordering $>_{sv}$ for resolution with the additional derivation rule (SAT1) preserves lifting in case of the Minimal Gödel Class. We have to show that whenever some literal L' in the ground instance C' of any clause C is allowed to be resolved upon, the corresponding literal L in C is also allowed to be resolved upon. For resolution without (SAT1) this does not hold because for some pairs of ground clauses $A\sigma$ and $B\sigma$ we have ($A\sigma >\!\!<_{sv} B\sigma$) (e.g, (P(a, a) $>\!\!<_{sv}$ P(a, a)), but there can be corresponding clauses A and B with variables such that (A $>_{sv}$ B) (eg, P(x,y) $>_{sv}$ P(y,y)) because of condition (2). Recall that the clause form of any formula in the Minimal Gödel Class does not contain constants. Thus the only literals in Herbrand expansion which do not contain functional terms with arguments, have a single new constant symbol (e.g, 'a') on all the argument places (e.g, S(a,a,a)). So, lifting is not preserved on these formulas only. Thus, whenever lifting is not preserved for some ground formula derived from Herbrand expansion, this ground formula is a substitution instance of a formula derived by (SAT1) for the variable-containing case. Now, the formula derived by (SAT1) "cures" lifting, as literals in it are not ordered.

An unpleasant consequence of using such a saturation rule is that it becomes unclear how to deal with subsumption. Obviously, we cannot use general subsumption, since then the clause derived by the saturation rule would be subsumed, as if it had never been derived. The following example demonstrates that full subsumption is not compatible with the last ordering. During the example we do not use (SAT1), since by full subsumption the derived clauses would be subsumed.

Consider the following unsatisfiable set of four clauses:

1: (P(x,y), R(x,x))
2: (¬R(x,y), P(x,x))
3: (R(x,y), ¬P(x,x))
4: (¬P(x,y), ¬R(x,x))

The $>_{sv}$-ordering allows to resolve upon the first literals only, using condition (2). At the first level we can only get the following two clauses, which both are tautologies.

1, 4 give 5: $(R(x,x), \neg R(x,x))$

2, 3 give 6: $(P(x,x), \neg P(x,x))$

Obviously we have to use these tautologies to get the refutation. On the second level we get:

5,2 give 7: $(P(x,x), \neg R(x,x))$

6,4 give 8: $(\neg P(x,x), \neg R(x,x))$

5,3 give 9: $(\neg P(x,x), R(x,x))$

6,1 give 10: $(P(x,x), R(x,x))$

Notice that all these are subsumed by clauses 1-4. Any clause derivable on the second level is obviously subsumed by clauses 1-4 since clauses 5 and 6 are tautologies. Thus we have to use the subsumed clauses in order to get the refutation:

7, 8 give 11: $\neg R(x,x)$

9, 10 give 12: $R(x,x)$

11, 12 give the empty clause.

As noted by N. Zamov, the use of saturation rule for any class decidable with the corresponding resolution variant can be postponed until no new clauses can be derived without saturation. The reason is that the amount of derivable clauses is finite (this follows from the proof of decidability).

4.4.2 Amending lifting by replacing clauses with certain instances

Whereas the method considered in the previous section consisted in using an additional derivation rule to derive certain instances of clauses, in [Joy76] the problem of amending lifting is overcome by replacing clauses with sets of instances. That is, in case some derived clause C fulfils certain properties, a new set of instances inst(C) is derived by a special saturation rule and C itself is deleted (compare also sections 5.3 and 5.4).

Any such a resolution variant using saturation combined with deletion of clauses needs a special completeness proof to show why the certain resolution variant is complete for a certain investigated class. The useful property of such resolution variants, when compared to variants decribed in the previous section, is that subsumption does not have to be restricted in any way.

Resolution variants of the "replacement" kind have been investigated by C. Fermüller (see [Fer91]), and are explained in detail in chapter 5 of the current monograph.

4.4.3 Amending lifting by defining a separate ordering for the ground case

N. Zamov has shown that although the ordering $>_{SV}$ cannot be lifted, the apriori refinement using this ordering is nevertheless complete, for Maslov's Class K, (which contains Gödel's Class) without any saturation (although this refinement won't be compatible with full subsumption). The proof of this fact can be inferred from the analogous proof for a very close ordering, given in chapter 6 dealing with Maslov's Class K.

The general method (originating from N. Zamov and V. Sharonov) for proving the completeness of the strategy based on some such ordering $>_0$ which cannot be lifted from the ground case, is to take some stronger ordering than $>_0$, say $>_0'$, for the ground case, such that $>_0'$ can be lifted to $>_0$. That is, the following modified version of condition (D) must hold for the input set and for all derived clauses and the corresponding Herbrand instances (not just ground instances of the input clauses):

(D') for any literals X, Y in some clause and corresponding Herbrand instances X' and Y' s.t. X', Y' occur in a ground clause derived from Herbrand expansion holds $(X >_0 Y) \Rightarrow (X' >_0' Y')$.

A good candidate for such an $>_0'$ is some non-syntactically defined ordering (eg, lock-resolution with the suitable indexing schema) for the ground case. Such a technique was originally used by N.Zamov for several completeness proofs, including the one in [Zam89].

The consequence of using a special underlying lock-resolution ordering is that full forward subsumption and tautology-elimination are not allowed. They are allowed only in cases correponding to π-ordering.

EXAMPLE 4.10.: For any t, P(x, y) is not allowed to subsume P(x, x){t/x}. For any t, the tautology {P(x, x), ¬P(x, x)}{t/x} cannot be removed. However, P(x, y) is allowed to subsume P(x, f(x, y)) and the tautology {P(x, y), ¬P(x, y)} may be removed.

4.5 THE CLASS E^+

In this section we will investigate some subclasses of the class called E^+. We will prove completeness of a certain refinement for a subclass of the class E^+. The ordering used here is much "less liftable" than the previously presented ordering $>_{sv}$, thus the method of using some strong lock-indexing for the ground case is somewhat more important.

In chapter 5 C. Fermüller investigates the same class E^+ using methods different from the current chapter: whereas we use a certain apriori refinement based on a generally non-liftable ordering, he uses an aposteriori refinement based on a certain A-ordering.

DEFINITION 4.5.: A functional term is called weakly covering iff for all subterms s of t that properly contain variables V(s) = V(t). An atom or literal A is called weakly covering iff each argument of A is either a ground term or a weakly covering term s.t. V(t) = V(A). (This definition appears in chapter 5).

For important properties of weakly covering terms and literals see chapter 5, section 5.1.3.

DEFINITION 4.6.: We say that a clause set S belongs to the Class E^+ iff for all clauses C in S the following holds:

(i) each literal in C is weakly covering.
(ii) for all literals L, M in C either V(L) = V(M) or $V(L) \cap V(M) = \emptyset$

The class just presented contains the clausal form of the class of formulas with function symbols where every atom contains only one variable; this class has been called N_1 and E (if given in terms of derivability, formulas in E can generally be assumed to be of the form $(\exists x) M$). Previously, the solvability of E has been shown in various ways. E contains, for example, the Skolemized version of Initially-extended Ackermann Class (the class of formulas without function symbols and with the prefix $(\exists x_1) \ldots (\exists x_n)(\forall y)(\exists z_1) \ldots (\exists z_m)$).

<u>EXAMPLE</u> 4.11.: The following clause set belongs to class E^+:

$\{P(f(f(a)), g(c, g(y, x))), \ P(u, f(u))\}$

$\{\neg P(f(h(x, y, z)), z), \ G(b), \ G(v), \ \neg G(u)\}$

$\{E(g(h(z, x, y), f(b)))\}$

$\{R(g(v, g(v, u)), g(u, v)), \ R(f(b), g(f(f(a)), g(u, v)))\}$.

In the paper [Tam 90] it is shown that in case the apriori v-refinement given below is complete for the Class E^+, it is also a decision strategy for E^+. However, so far we have not been able to prove the completeness of this refinement for E^+, nor have we found any counterexamples to this.

As to the decidability of E^+, as shown by C. Fermüller in chapter 5, a simple saturation rule added to the v-refinement will guarantee completeness of the procedure. Then a simple modification of the theorem 4 below gives decidability of E^+. However, from the proof-search aspect the saturation rule is very unwelcome, and we would still like to find a provably complete resolution refinement without having to use saturation.

In the following we will give a completeness proof of the apriori v-refinement for the clause form of the class E, which, as previously noted, is a subclass of E^+.

4.5.1 Completeness of a v-refinement for the class E

<u>DEFINITION</u> 4.7.: Let A and B be literals. We say that $A >_v B$ iff the following is true:

$V(A) = V(B)$ and for all y in $V(A)$ holds $\tau_{max}(y, B) < \tau_{max}(y, A)$.

The v-refinement is defined by the apriori restriction on a resolution rule: only the maximal (in the sense of new ordering $>_v$) literals in a clause are allowed to be resolved upon.

EXAMPLE 4.12.: In the clause A(f(f(y,x),y),a), B(b,g(x,y)), C(y,b,x) only the literal A(f(f(y,x),y),a) is allowed to be resolved upon.

The proof presented in [Tam90] and in section 4.5.3 of the fact that only a finite number of clauses can be derived by the v-refinement from any clause set in the Class E^+, has the following structure: it is shown that the v-refinement has this property for the Class $E^{+'}$, which is the same as the Class E^+, except for the property (ii), which in $E^{+'}$ is replaced by a property (ii'):

(ii') for all literals L, M in C holds V(L) = V(M).

If the v-refinement decides $E^{+'}$, it is clear that v- refinement with splitting decides the class E^+, since every formula in E^+ can be split into a finite set of formulas in $E^{+'}$.

Notice that instead of using splitting, the decision algorithm for E^+ is obtained from the same algorithm for $E^{+'}$ by either using an aposteriori refinement of Robinson's resolution or using a corresponding restriction on factorization.

4.5.2 Completeness proof of the v-refinement for a class E'

DEFINITION 4.8.: We say that a clause set S belongs to the Class E' iff the following properties hold for all clauses C in S:

1. Each literal in C is weakly covering.
2. Either C is ground or V(C) = {x} for some variable x and for each literal L in C holds V(L) = {x}.

We will now present a proof of the completeness of the v- refinement for the class E'. The general completeness of this refinement is still an open problem.

The structure of the proof: first we define a certain indexing for all literals in the ground instances of the given clauses. Then, given that the set of ground clauses is unsatisfiable, we can find a lock-resolution proof with the defined indices. We will show that the ordering of literals in the ground clauses given by indices can be lifted to the $>_v$ -ordering for clauses with variables; the lifting is immediate for the original set of clauses and can be shown to hold for all new clauses generated by resolution with the refinement from a clause set in Class E'.

Indexing

Let S be a clause set from the Class E'. Consider the Herbrand expansion H of S: Each clause C' in H is an instance of some clause C from S: C' = C{t/x} where x is a single variable in C, t is a ground term from the Herbrand universe of S. We will give each occurrence of the literal $L_j = L_j\{t/x\}$ in C' the following index: $\tau(t) + \tau_v(L_j)$, where $\tau_v(L)$ is the depth of the deepest occurence of a variable in L.

EXAMPLE 4.13.: The following is a sample indexing:

P(x, ffa)	P(fa, x)	P(fa, fx)	
P(a,ffa) :0	P(fa,a) :0	P(fa,fa) :1	a/x
P(fa,ffa) :1	P(fa,fa) :1	P(fa,ffa) :2	fa/x
P(ffa,ffa) :2	P(fa,ffa) :2	P(fa,fffa) :3	ffa/x
P(fffa,ffa) :3	P(fa,fffa):3	P(fa,ffffa):4	fffa/x
...	...	...	...

As we can see, syntactically equal literals in the Herbrand expansion can have different indexes, since the index of an occurrence of a literal in the expansion depends on a corresponding literal in the unexpanded clause set.

LEMMA 4.1.: Let (L_1', ..., L_n') and (R_1', ..., R_m') be two clauses in the Herbrand expansion of S, corresponding to (L_1, ..., L_n) and (R_1, ..., R_m) in S. Let $L_1' = R'^d_1$ Then, if the clause C = $(L_2, \ldots, L_n, R_2, \ldots, R_m)\sigma$, where σ = m.g.u(L_1, R^d_1) is not ground, the index of L_1' must be the same as the index of R_1'.

Proof:

Suppose that C is not ground. Then, due to property (B), both L_1 and R_1 contain variables. Let x be the variable in L_1, y be the variable in R_1. Consider the substitution σ = m.g.u(L_1, R_1^d). Since C is not ground, there must be at least one variable (either x or y) which is not bound to a ground term by σ . Suppose that x is a variable which is not bound to a ground term. Then we have three possibilities for binding of x in σ :

(1) y/x. Then the maximal depth of x is the same as the maximal depth of y and, since $L_1'=R_1'$, the term substituted for x in L_1' is the same as the term substituted for y in R_1'.

(2) f[y]/x, where f[y] is some term containing y in R_1. Then the depth of the term substituted for x in L_1' is the same as the depth of the term substituted for variable y in R_1' minus the maximal depth of y in f[y].

(3) g[x]/x, where g[x] contains x. This cannot happen by definition of a m.g.u.

Thus we get that the index of L_1' must be the same as the index of R_1'.

Q.E.D.

THEOREM 4.3.: The v-refinement of resolution is complete for E'.

Proof:

Induction base: Lifting of the given indexing ordering to the $>_V$-ordering for the originally given set of clauses is immediate.

Induction step: Let lifting of the ordering hold for non-ground clauses A and B. Let resolution generate a clause C from clauses A and B. If C is ground, we can, for example, use splitting (or some restrictions on C based on splitting). So the important case is where C is not ground. From lemma 1 we get that for any ground instances A' and B' in the Herbrand expansion the literals resolved upon have the same index. Then it is easy to see that ordering of the clause C', given by resolution from A' and B' is also lifted, that is, corresponds exactly to the $>_V$-ordering of a non-ground clause C.

Q.E.D.

As we proved completeness by using a lock-strategy for the ground case, we cannot eliminate tautologies and cannot use unrestricted subsumption, since lock-strategy is incompatible with these. However, we can do partial subsumption, where the ordering of the literals in the clause is also taken into consideration.

<u>EXAMPLE</u> 4.13.: Consider the following refutable set of four clauses from E':

1: (P(f(a),f(x)), R(x,f(f(a))))

2: (¬ R(a,f(x)), P(x,x))

3: (R(a,f(x)), ¬P(x,x))

4: (¬P(f(a),f(x)), ¬R(x,f(f(a))))

By the $>_v$-ordering only the first literals in these four clauses are allowed to be resolved upon. At the first level we can derive two new clauses, both of them tautologies:

1,4 give 5: (R(x, f(f(a))), ¬R(x, f(f(a))))

2,3 give 6: (P(x, x), ¬P(x, x))

The clauses derived at the next level are subsumed by clauses 1-4. However, when subsumption is restricted by the ordering, these clauses are not subsumed and the refutation follows easily.

4.5.3 Termination of the v-refinement for the Class $E^{+'}$

The following presents a proof that the v-refinement terminates on any clause set from the Class $E^{+'}$. Although the completeness of the v-refinement for Class $E^{+'}$ is still an open problem, we decided to present the termination proof for the following reasons: first, the techniques used in the proof could be useful elsewhere; second, since we have not found any counterexamples even to the general completeness of the v-refinement, it could be the case that v-refinement is indeed complete, and should the completeness proof be found in the future, the following termination proof can be put to an immediate use. Lemmas 4.2, 4.3 and 4.4 are analogous to the subcases of the lemma 5.5 in chapter 5.

LEMMA 4.2.: If clauses D_1 and D_2 both have the property (ii') (in the sense of the definition of Class E^{+}') and every literal in D_1 and D_2 is weakly covering, then all clauses inferred from D_1 and D_2 using v-refinement also have the property (ii') and every literal in them is weakly covering.

Proof:

1) Property (ii'): If (ii') holds for C, then (ii') holds for $C\sigma$, where C is any clause and σ is any substitution. If the clause C is obtained from clauses D_1 and D_2 with property (ii') by v-refinement, resolving upon literal L_1 in D_1 and literal L_2 in D_2, then L_1 contains all variables in D_1 and L_2 contains all variables in D_2.

2) Weakly covering: Obviously factorization preserves weakly covering (the value of the m.g.u, used for factorizing D_1, cannot have elements of the kind t/x, where t contains variables, since then by weakly covering and (ii') t must also contain x). Next we will consider a binary resolution step.

In order that weakly covering were not preserved, such a substitution σ must be applied to some literal L in D_1 (let $\bar{x} = \{x_1, \dots, x_k\} = V(L)$) that σ contains t/x_i $(1 \leq i \leq k)$, t contains variables and L contains some variable $x_j \neq x_i$ such that $L\sigma$ also contains x_j and t does not contain x_j. $\sigma = \text{m.g.u}(L_1, L_2^d)$, where by property (ii') and v - refinement L_1 contains variables $x_1, \dots, x_k$. Let $y = \{y_1, \dots, y_r\} = V(L_2)$.

Suppose that σ contains $t[\bar{y}']/x_i$ $(1 \leq i \leq k)$ (where $\bar{y}'$ is a subset of variables in L_2) and L contains a variable $x_j = x_i$. We will examine whether $L\sigma$ can contain x_j. For this we will investigate possible cases for some term d in L_2 which corresponds to some occurrence of x_j in L_1.

Consider the following cases:

- x_j occurs in some term $g[\bar{x}]$ in L_1'. This cannot happen by definition of a m.g.u.
- $d = y_n$ for some y_n in $\bar{y}$. Then either $t\sigma$ contains x_j or $L\sigma$ does not contain x_j.
- d is a ground term. Then $L\sigma$ does not contain x_j.
- $d = g[\bar{y}]$ for some non-ground term $g[\bar{y}]$. Then $L\sigma$ does not contain x_j.

Q.E.D.

LEMMA 4.3.: If clauses D_1 and D_2 both have the property (ii') and every literal in them is weakly covering, then for any clause D inferred from D_1 and D_2 with a factorization or binary resolution step of v-refinement $\tau_v(D) < \max(\tau_v(D_1), \tau_v(D_2))$.

Proof:

Let D be a clause inferred from D_1 and D_2 using the v- refinement. D can contain some variable deeper than any variable in D_1 and D_2 only in case a substitution used for inferring D contains such an element $t[\bar{y}]/x_i$ (where $\bar{y}$ is some set of variables) that some literal $L\sigma$ in D is such that x_i occurs in L (L is the original of $L\sigma$ in D_1 or D_2) in some term g and σ does not bind all variables in y to ground terms. We will show that this is impossible.

1. Inference by factorization. Since (ii') holds for D_1 and every literal in D_1 is weakly covering, factorization cannot give such a substitution σ that contained $t[\bar{y}]/x_i$ (then $\bar{y}$ should contain x_i, which is impossible).

2. Inference by resolution. Let $L, L_1 \in D_1$, $L_2 \in D_2$.
   $V(D_1) = \{x_1, \ldots, x_k\}$, $V(D_2) = \{y_1, \ldots, y_r\}$.
   Let $\sigma = \text{m.g.u}(L_1, L_2^d)$.

Since we use the v- refinement, the variable x_i cannot occur deeper in L than in L_1. Let $f[x_i]$ be the term in L_1 where x_i has its deepest occurence in L_1. We can assume that D_2 does not contain variables deeper than D_1. Due to the v-refinement, $f[x_i]$ has the deepest occurrences of variables in D_1, D_2.

Suppose that σ contains the element $t[\bar{y}]/x_i$. Then during the computation of σ the term $f[x_i]\{t[\bar{y}]/x_i\} = f[t[\bar{y}]]$ must have been unified with some term d in L_2. Consider the following cases:

1) d is a variable. Then d cannot be unified with $f[t[\bar{y}]]$, since $\bar{y}$ contains d.

2) d is a ground term. Then every variable in $\bar{y}$ is bound to some ground term, thus σ cannot contain an element $t[\bar{y}]/x_i$ s.t. variables in $\bar{y}$ were not bound to some ground term.

3) $d = f[g]$, g is either a variable or a ground term and g occurs in the term $f[g]$ on the same place as $t[\bar{y}]$ first occurs in $f[t[\bar{y}]]$. If g is a variable, then d cannot be unified with $f[t[\bar{y}]]$ since $\bar{y}$ contains g. If g is a ground term, then every variable in $\bar{y}$ is bound to some ground term, thus σ cannot contain an element $t[\bar{y}]/x_i$ s.t. variables in $\bar{y}$ were not bound to some ground term.

4) $d = f[t'[\bar{y}]]$, where $t'[\bar{y}]$ occurs in $f[t'[\bar{y}]]$ on the same place as $t[\bar{y}]$ first occurs in $f[t[\bar{y}]]$. Suppose that $t[\bar{y}]$ and $t'[\bar{y}]$ are unifiable. In case $t[\bar{y}]$ contains non-ground functional terms as proper subterms, then iterate our analysis (notice that if some literal L is weakly covering, then any subterm of L is weakly covering) until such a term $r[\bar{y}]$ is reached which does not contain non-ground functional terms as proper subterms.

Therefore, in order that σ existed and $L\sigma$ contained some variables in $\bar{y}$, L_2 must contain such a term $f[t'[\bar{y}]]$ that the maximal depth of variables in $f[t'[\bar{y}]]$ were not smaller than the maximal depth of variables in $f[t[\bar{y}]]$. Since $f[x_i]$ is assumed to contain the deepest occurrence of variables in D_1 and D_2, $f[t[\bar{y}]]$ contains variables deeper than the maximal depth of variables in D_1 and D_2, which is impossible. Thus no variable in $D\sigma$ may occur deeper than the depth of the deepest variable in D_1, D_2.

Q.E.D.

Corollary 4.1.: v-refinement of the resolution method cannot infer a clause D from the clause set S in class E+' such that any variable occurred deeper in D than the maximal depth of variables in S.

Proof:
immediate from lemmas 3 and 4.

Q.E.D.

Let S be some clause set in Class $E^{+'}$. Let T be a set of all non-ground functional terms in S (if T is empty, the proof of the forthcoming theorem 4.4 is trivial). Let L be some literal in S. Since T is a finite set, there must be a finite set L^* of such literals which can be obtained from L by applying substitutions of the kind $\{t_1/x_1, \dots, t_k/x_k\}$ (where $\bar{x} = \{x_1, \dots, x_k\}$ is some set of variables, and t_i denote arbitrary elements of T) to L so that no variable occurs deeper in L^* than $\tau_v(S)$. (T and L^* are defined up to renaming the variables).

EXAMPLE 4.14.: $S = \{\{A(f(x)), L(x,f(a))\}, \{\neg P(f(g(u,y)),u)\}\}$
Then $\tau_v(S) = 2$, $T = \{f(x), f(g(u,y)), g(u,y)\}$.

If $L = A(f(x))$, then

$L^* = \{A(f(x)), A(f(g(x,y))), A(f(f(x)))\}$. If $L = \neg P(f(g(u,y)), u)$, then

$L^* = \{\neg P(f(g(u,y)),u)\}$.

DEFINITION 4.9.: Let L be some literal such that we can compute L^*. VNmax(L) denotes the maximal number of different variables in literals in L^*.

EXAMPLE 4.15.: Let S be the same as in the previous example.

$$\begin{aligned} VNmax(A(f(x)) &= 2, \\ VNmax(A(f(f(x))) &= 1, \\ VNmax(A(f(a))) &= 0, \\ VNmax(\neg P(f(g(u,y)), u)) &= 2, \\ VNmax(A(f(f(a)))) &= 0, \\ VNmax(A(x)) &= 4 \ \{\text{Notice } A(g(g(u'', y''), g(u', y')))\}. \end{aligned}$$

LEMMA 4.4.: Let S be a clause set in E+'. The v-refinement of resolution cannot infer a clause $(L_1', \dots, L_n')$ from S such that for some $L_i'(1 < i < n)$; $VNmax(L_i') > VNmax(L_i)$, where L_i is the original of L_i'.

Proof:
Immediate from corollary 1 and the definition of VNmax.

Q.E.D.

LEMMA 4.5.: Let clauses D_1 and D_2 have the property (ii') and each literal in D_1, D_2 be weakly covering. Let the v-refinement infer a new clause $D = (L_1', \dots, L_n')$ from D_1 and D_2 such that D contains a term deeper than the deepest term in

D_1, D_2. Then:

$$\forall_i (1 < i < n) \begin{cases} VNmax(L_i') = 0 \text{ if } VNmax(L_i) = 0 \\ VNmax(L_i') < VNmax(L_i) \text{ if } VNmax(L_i) \neq 0 \end{cases}$$

where L_i the original of L_i' in D_1 or D_2, VNmax is computed in respect to the original formula in E+' from which D_1, D_2 were inferred.

Proof:

Due to lemma 4.4 the only possibility for some L_i' to contain a term deeper than the deepest term in D_1 and D_2 is to substitute some ground term t for such a variable x_i in L_i such that x_i occurs in some functional term $f[x_i]$ in L_i. Then x_i will occur in no term in D, thus the VNmax of all literals in D_1 which contain variables (VNmax = 0) is bigger than VNmax of the corresponding literals in D.

Next we will consider the clause D_2, which did not contain x_i. Let x_i have a deepest occurrence in L_1 in a term $g[x_i]$. Notice that after applying a substitution $\{t/x_i\}$ to the literal L_1 in D_1 (assume L_1 is resolved upon in D_1) the term $g[x_i]$ will turn into $g[t]$, which by assumption is deeper than the deepest one in D_1, D_2 (since we use v-refinement, x_i has a deepest occurrence in L_1) . Let d be the term which occurs in R_1 (assume R_1 is resolved upon in D_2) on the place of the first occurence of $g[x_i]$ in L_1. Consider the following cases:

- $d = g[t]$. This is impossible, since $g[t]$ is deeper than the deepest term in D_1, D_2.

- $d = g[y_j]$. Then literals in D which correspond to literals in D_2 will contain a ground term on place of a variable y_j, thus VNmax of these literals in D will be less than VNmax of their originals in D_2.

- d is a variable. Then literals in D which correspond to literals in D_2 will contain a ground term on place of a variable d, thus VNmax of these literals in D will be less than VNmax of their originals in D_2.

Q.E.D.

<u>THEOREM</u> 4.4.: The v-refinement of the resolutiom method infers a finite number of clauses from any clause set in the Class $E^{+'}$.

<u>Proof</u>:
Let S be a clause set in the class $E^{+'}$. Due to corollary 4.1 no clause inferred from S with the v-refinement method with V-strategy can contain variables deeper than $\tau_V(S)$. All literals in S have a certain VNmax. Lemmas 4.4 and 4.5 demonstrate that if a depth of a literal becomes bigger than a certain given bound for S, then VNmax of a literal will be correspondingly smaller. By lemma 4.4 VNmax of a literal cannot grow. Thus such a bound on the depth of terms must exist that all literals of this depth have a VNmax equal to 0, thus they do not contain variables and their depth cannot grow.

Therefore by v-refinement one cannot infer literals deeper than the before-mentioned depth bound. Thus the number of inferrable literals (up to renaming variables) is also bounded. Then, due to lemma 4.3, the length of the inferrable clauses is bounded and thus the number of inferrable clauses.

Q.E.D.

Chapter 5

SEMANTIC TREE BASED RESOLUTION VARIANTS

5.1 PRELIMINARIES

In this chapter we shall consider an approach to the decision problem which is basically the one worked out by William H. Joyner in his dissertation [Joy 73] (see also [Joy 76]). Most of the results of this chapter already appeared in [Fer91a] and [Fer91]. We shall use A-ordering strategies and saturation mechanisms to guarantee that there are only finitely many resolvents for classes of clause sets. A kind of uniform frame for the results of this chapter is provided by the fact that all employed variants of resolution are demonstrated to be complete (at least for certain classes of clause sets) via semantic trees. Unlike in the cases investigated by Joyner, the depth of a resolvent will generally not be bounded by the maximum of the term depth of its parents. Anyway, it suffices to show that a "global" term depth limit exists, i.e. the term depth of resolvents cannot grow arbitrarily often. In order to achieve a decision procedure we also have to guarantee a length limit for the generated clauses. This, in most of the cases described below, will be achieved by employing the splitting rule.

Our reinvestigation of A-orderings and saturation should make clear that the method is by no means restricted to the prefix classes that have been considered by Joyner. On the contrary, to rely on the flexibility of the clause syntax terminology also for the definition of the decision classes, allows for both: generalisations of Joyner's results and more transparent procedures and proofs.

A-orderings can be described as a special subcase of π - orderings as treated in chapter 4. Nevertheless, from a methodological point of view, it seems worthwhile to study A-ordering strategies indepedently of this more general context of π-ordering refinements: A- orderings are purely syntactically defined π- orderings on atoms (not literals!), which e.g. means that no reference to the location of clauses within a "proof tree" is needed. Therefore they are somewhat easier to describe and handle than general π - orderings. (See chapter 4 for more details). More importantly, we may employ the well known device of semantic trees for the completeness proofs, which directly implies the admissibility of all types of subsumption and tautology elimination rules for the resolution variants of this chapter.

5.1.1 A-orderings

The concept of A-orderings was introduced by Kowalski and Hayes [KH69]. We present a definition, that can easily be demonstrated to be equivalent with that of Kowalski and Hayes.

DEFINITION 5.1.: An A-ordering $<_A$ is a binary relation on atoms s.t.

(i) $<_A$ is irreflexive,

(ii) $<_A$ is transitive,

(iii) for all atoms A, B and all substitutions ϑ:

$$A <_A B \text{ implies } A\vartheta <_A B\vartheta.$$

Joyner remarks that this definition should be augmented by the following condition:

(iv) for every clause set S, $<_A$ is compatible with some enumeration of the corresponding Herbrand base (see def. 2.31). I.e. there is some enumeration $A_0, A_1 \ldots$ of all ground atoms that are Herbrand instances of atoms occurring in S s.t. $A_i <_A A_j$ implies $i < j$.

According to Joyner, Kowalski and Hayes make implicit use of such an ordering compatible enumeration in their completeness proof. Thus, in Joyner's version of A-orderings, it is e.g. not allowed to order all atoms starting with predicate letter "P" in front of all atoms starting with "Q", although this ordering trivially fulfills conditions (i) - (iii). But anyway, we shall see in section 5.1.2 that condition (iv) is not necessary to ensure the completeness of the resulting resolution strategy.

A more subtle point concerns the definition of the resolvents that obey the A-ordering restriction. Whereas Kowalski and Hayes demand the atoms to be resolved upon to be maximal w.r.t. the ordering, we shall use the following definition which is due to Joyner.

DEFINITION 5.2.: For any clause set S and any A-ordering $<_A$:

$E \in R_{<_A}(S)$ iff E is a Robinson-resolvent of clauses in S s.t. for no atom B in E: $A <_A B$, where A is the resolved atom.

According to the terminology developed in chapter 4 this means that we are employing an a posteriori criterion as opposed to the weaker (i.e. less restrictive) priori criterion as given by Kowalski and Hayes.

It is interesting to mention that in order to arrive at decision procedures for certain classes of clause sets we have to insist on the stronger a posteriori condition.

5.1.2 Completeness via semantic trees

In the sections to come we shall not only refer to the completeness of A-ordering strategies, but also have to present completeness proofs for some more special resolution refinements. For this purpose we rely on the the important device of semantic trees. To remain fairly selfcontained and to easify the argumentation in later sections we present a proof of the (refutational) completeness of A-ordering strategies. We stress that the definition and theorems in this section are not new; versions of them rather can be found e.g. in [KH69] or [Joy73].

Semantic trees have been introduced by Robinson [Rob68] and remained a standard tool in the field of automated theorem proving ever since. Our version of semantic trees is basically the one of Kowalski and Hayes [KH68]. We repeat the relevant definitions, but assume the reader to be familiar with the basic vocabulary of graph theory.

DEFINITION 5.3.: A semantic tree based on a set of atoms K is a binary tree with elements of K and their negations labelling the edges in the following way:
Let A_0, A_1,... be an enumeration of K. The two edges leaving the root of the corresponding semantic tree T are labelled A_0 and $\neg A_0$, respectively; and if either A_i or $\neg A_i$ labels the edge entering a node, then A_{i+1} and $\neg A_{i+1}$ label the edges leaving that node. With each node of T we associate a refutation set, consisting of the literals labelling the edges of the path from the root down to this node.

We shall only use semantic trees that are based on finite subsets of some Herbrand base (for the relevant definitions we refer to chapter 2).

DEFINITION 5.4.: A clause C fails at a node of a semantic tree T iff the complement of every literal in some ground instance $C\gamma$ of C is in the refutation set of that node. A node n is a failure node for a clause set S iff some clause of S fails at n but no clause of S fails at a node above n (i.e. a node with a shorter path to the root). A node is called inference node iff both of its sons are failure nodes. T is closed for S iff there is a failure node for S on every branch of T.

The completeness proofs of this chapter essentially rest on the following well known corollary to Herbrand's Theorem:

THEOREM 5.1.: If a clause set S is unsatisfiable then there is a finite subset K of $\hat{H}_S$ s.t. every semantic tree T based on K is closed for S.

A proof of theorem 5.1 can e.g. be found in [KH69].

Resolution strategies R are usually defined such that, for any clause set S, the set of resolvents R(S) is a proper subset of the set of Robinson-resolvents. In contrast, we shall also investigate strategies where R(S) contains certain instances of Robinson-resolvents.

DEFINITION 5.5.: Let C be a class of clause sets. We call a resolution strategy R complete via semantic trees on C iff

(i) for any unsatisfiable S in C there is some finite subset K of $\hat{H}_S$ s.t. there exists a semantic tree T based on K closed for S,

(ii) for any two clauses C, D which fail immediately below an inference node of T there is an R - resolvent of C and D which fails at the inference node.

Given theorem 5.1 it is an easy task to prove

LEMMA 5.2.: Any resolution strategy R which is complete via semantic trees on some class C is a (refutational) complete deduction procedure for class C.

Proof: Let S be an unsatisfiable clause set in C. Observe that, theorem 5.1 guarantees the existence of a semantic tree T as specified in condition (i) of definition 5.5. It remains to show that condition (ii) implies Box in $R^*(S)$.

Q.E.D.

Because $S \subseteq R(S)$ T is also closed for R(S) and any failure node for S is either also a failure node for R(S) or some predecessor of this node is a failure node for R(S). Since T is closed for S there must be at least one inference node for S in T. But then, by condition (ii), some failure node for R(S) is a predecessor of at least two failure nodes for S. This means that the number of failure nodes shrinks strictly monotonically as we iteratively apply R. Therefore there exists some n s.t. some clause in $R^n(S)$ fails at the root of T. But the only clause failing at the root of a semantic tree is $\Box$.

Q.E.D.

We are now prepared to present a completeness proof for A - ordering strategies.

THEOREM 5.3: For any A - ordering $<_A$ $R_{<_A}$ is a resolution strategy which is complete on all classes of clause sets.

Proof:

Let S be any unsatisfiable clause set. By theorem 5.1 there is a finite $K \subseteq \hat{H}_S$ s.t. any semantic tree T for K is closed for S. Because of lemma 5.2 it suffices to show that we can choose T s.t. if clauses C and D define an inference node

in T then some $R_{<_A}$ - resolvent E of C and D fails above this node.
To this aim we order the atoms of K in a list $A_0, A_1, \ldots, A_n$ compatible with $<_A$; i.e. $A_i <_A A_j$ implies $i < j$. Let T be the semantic tree labelled according to this enumeration of K (compare definition 5.3). Let C and D be clauses in S failing immediately below an inference node n of T. Let A_k and $\neg A_k$ be the two ground literals labeling the edges leaving the inference node n. Clearly, there is a Robinson- resolvent E of C and D s.t. A_k is ground instance of the corresponding resolved atom A. We show that E also is an $R_{<_A}$ - resolvent.

Let γ be the substitution s.t. $E\gamma$ is in the refutation set of n. By definition of T we have $A\gamma \not<_A B\gamma$ for all atoms B in E. But since $<_A$ is an A- ordering $A\vartheta \not<_A B\vartheta$ implies $A \not<_A B$ for all substitutions ϑ. Therefore E fulfills the defining conditions of an $R_{<_A}$- resolvent.

Q.E.D.

A look at the terminology used in the above proof reveals why, contrary to the comments of Joyner, condition (iv) as refinement of the definition of A - orderings is empty: As we based the semantic trees on finite subsets of the Herbrand base there is always an enumeration of the relevant atoms which is compatible with the ordering.

As for the completeness in presence of the splitting rule (see chapter 2), note that if a clause C fails at some node n then clearly all split components of C fail at n, too. Thus all resolution strategies that are complete via semantic trees remain complete if combined with the splitting rule. Similar facts hold for the usual deletion strategies: A tautology does not fail at any node of a semantic tree, because no refutation set can contain both, an atom and its negation. Moreover if C subsumes D, i.e. there is a substitution ϑ s.t. $C\vartheta \subseteq D$, then clearly C fails at any node at that D fails. Therefore tautologies and subsumed clauses may be removed from the sets of resolvents without effecting completeness (for the details we again refer to [KH69]).
We remark that actually it is not necessary to employ the - computationally complex - splitting rule in order to get decision procedures. However, the proofs are much simpler if the clauses are assumed to be decomposed.

5.1.3 Covering terms and atoms

Throughout the next sections we speak of terms, atoms and literals characterized by special properties which we shall call covering and weakly covering, respectively:

DEFINITION 5.6.: A functional term t is called covering iff for all functional subterms s occurring in t we have V(s) = V(t). An atom or literal A is called covering iff each argument of A is either a constant or a variable or a covering term t s.t. V(t) = V(A).

DEFINITION 5.7.: A functional term t is called weakly covering iff for all non ground, functional subterms s of t: V(s) = V(t). Similary to definition 5.6, an atom or literal A is called weakly covering iff each argument of A is either a ground term or a variable or a weakly covering term t s.t. V(t) = V(A).

EXAMPLES: g(h(x), a, f(x, x)) is a covering term f(x, g(h(a), f(y,x), y)) is weakly covering but not covering; g(f(x, y), x, h(x)) is not (weakly) covering. P(h(x), x) is a covering and Q(f(x, f(x,y)), f(y, x), h(a)) a weakly covering atom; P(f(a, h(b)), h(c)), like any ground atom, is covering, too. P(h(x), y) and P(h(x), f(x,y)) are not weakly covering.

Clearly, all covering atoms or terms are also weakly covering, but the converse does not hold. Covering atoms originate for instance by Skolemization of (pure) prenex formulas when the universal quantifiers precede all existential quantifiers. Also observe that any atom or term that contains at most one variable is weakly covering.

In section 5.4.1 below we shall prove that neither the term depth nor the number of variables of the resolvents in a certain class of clause sets, which is defined in terms of covering literals, can increase. For this purpose we need the following lemma:

<u>LEMMA</u> 5. 4: Let ϑ be a m.g.u. of two covering atoms A and B; then the following properties hold for $A\vartheta$ (= $B\vartheta$):

(i) $A\vartheta$ is covering,

(ii) $\tau(A\vartheta) = \max(\tau(A), \tau(B))$

(iii) $|V(A\vartheta)| \leq \max(|V(A)|, |V(B)|)$.

<u>Proof</u>:

Some additional terminology will result in a more concise formulation of the proof: Let us write $A \geq B$ if A = B or if B can be obtained from A by substituting some variables of A by constants or other variables. Observe that,

(*) if $A \geq B$, then B is covering whenever A is covering.

Moreover we write $A < B$ if each functional subterm t of B contains all variables of A (i.e. $V(A) \subseteq V(t)$ for all functional subterms t of B).

We have to trace the process of unification. Let ρ_i denote the concatenation of the mesh substituents applied so far, i.e.

$\rho_0 = \varepsilon$,

$\rho_i = \{ t_1/x_1 \} \cdot \{ t_2/x_2 \} \cdot \ldots \{t_i/ x_i \}$, where

$\{t_i / x_i\}$ denotes the i-th mesh substituent. (See chapter 2. 5) .

We prove that $[A \geq A\vartheta$ or $B \geq B\vartheta]$ by induction in the number of substitution steps using the following induction hypothesis:

Either

(IH1) $A \geq A\rho_i$ and $B \geq B\rho_i$ or

(IH2) $A \geq A\rho_i$ and $A\rho_i < B\rho_i$ or

(IH3) $B \geq B\rho_i$ and $B\rho_i < A\rho_i$

(IH1) trivially holds for $i = 0$. Let $\{ t_{i+1}/ x_{i+1} \}$ be the next mesh substituent. W.l.o.g. we may assume that x_{i+1} is found in $A\rho_i$ and t_{i+1} in $B\rho_i$ in the point of disagreement between $A\rho_i$ and $B\rho_i$ (otherwise exchange $A\rho_i$ and $B\rho_i$) .

We have to consider the following cases:

(1) (IH1) holds:

(1a) t_{i+1} is a variable or a constant:
This means that $A\rho_i \geq A\rho_{i+1}$ and $B\rho_i \geq B\rho_{i+1}$. Thus, by transitivity of the relation " $\geq$ ", (IH1) also holds for i+1.

(1b) t_{i+1} is a functional term:
By (IH1) and (*) $B\rho_i$ is covering. Therefore $V(t_{i+1}) = V(B\rho_i)$. This implies that $B\rho_i < A\rho_{i+1}$ (= $A\rho_i \cdot \{t_{i+1}/x_{i+1}\}$). Now observe that x_{i+1} is not in t_{i+1} (otherwise A and B would not be unifiable). But as $B\rho_i$ is covering this amounts to $B\rho_i = B\rho_{i+1}$ Summarizing we have shown that, in this case, (IH3) must hold for i+1.

(2) (IH2) holds:

(2a) t_{i+1} is a variable or a constant: Clearly $A\rho_i \geq A\rho_{i+1}$ and $A\rho_{i+1} < B\rho_{i+1}$. Therefore, similiar to case (1a), (IH2) remains valid.

(2b) t_{i+1} is a functional term: This cannot happen because $A\rho_i < B\rho_i$ implies $V(A\rho_i) \subseteq V(t_{i+1})$. This, in turn, would mean that x_{i+1} in $V(t_{i+1})$ which contradicts the assumption that A and B are unifiable.

(3) (IH3) holds: The argument is completely analogous to case (2).

Summarizing we have shown that $A \geq A\vartheta$ or $B \geq B\vartheta$. As A and B are covering it follows that $A\vartheta$ (= $B\vartheta$) is covering, too. Moreover, by definition of $\geq$, $A \geq A\vartheta$ implies $\tau(A) = \tau(A\vartheta)$, therefore $\tau(A\vartheta) = \max(\tau(A), \tau(B))$. Finally, observe that $A\vartheta$ cannot contain more than $\max(|V(A)|, |V(B)|)$ variables.

Q.E.D.

To assist the formulation of a similiar lemma for weakly covering atoms we introduce the following notation:

<u>DEFINITION</u> 5.8.: For any atom or literal A let $\Gamma(A)$ be the set of all ground terms that are subterms of A. Additionally, let $\tau_g(A)$ be the term depth of the deepest term in $\Gamma(A)$; i.e. $\tau_g(A) = \max\{\tau(t) \mid t \text{ in } \Gamma(A)\}$. The definitions of Γ and τ_g are extended to clauses C in the obvious way, i.e. $\Gamma(C) = \bigcup_{L \in C} \Gamma(L)$ and $\tau_g(C) = \max\{\tau_g(L) \mid L \text{ in } C\}$.

DEFINITION 5.9.: For any atom, literal or clause E let $\tau_v(E)$ be the maximum of $\tau_{max}(x, E)$ over all $x \in V(E)$. If E is ground we define $\tau_v(E) = 0$.

Observe that in any covering or weakly covering atom the maximal depth of occurrence is the same for all variables of the atom. Thus $\tau_v(A) = \tau_{max}(x, A)$ for every $x \in V(A)$.

We may now state:

LEMMA 5.5.: Let ϑ be a m.g.u. of two weakly covering atoms A and B then the following properties hold for $A\vartheta$ (= $B\vartheta$):

(i) $A\vartheta$ is weakly covering,

(ii) $\tau_v(A\vartheta) \leq \max(\tau_v(A), \tau_v(B))$,

(iii) $\tau(A\vartheta) \leq \max(\tau_v(A), \tau_v(B)) + \max(\tau_g(A), \tau_g(B))$,

(iv) either $\Gamma(A\vartheta) \subseteq \Gamma(A) \cup \Gamma(B)$ or $A\vartheta$ is ground,

(v) $|V(A\vartheta)| \leq \max(|V(A)|, |V(B)|)$.

Proof:

We proceed similarly to the proof of lemma 5.4. For atoms A, B and a set of ground terms G let us write $A \geq_G B$ if $A = B$ or if B can be obtained from A by substituting some variables of A by other variables or some member of G. It clearly holds that,

(*') if $A \geq_G B$, then B is weakly covering whenever A is weakly covering (for any set of ground terms G).

Moreover we write $A <' B$ iff each subterm t of B, that properly contains variables, contains all variables of A (i.e., $|V(t)| = 0$ or $V(A) \subseteq V(t)$ for all subterms t of B).

Again, let ρ_i denote the concatenation of the first i mesh substituents of the unifier.

In the following we use E as an abbreviation for $\Gamma(A) \cup \Gamma(B)$ and prove that $A \geq_E A\vartheta$ or $B \geq_E B\vartheta$ by induction on the number of mesh substitions. The induction hypothesis now reads:

Either

(IH1)	$A \geq_E A\rho_i$	and	$B \geq_E B\rho_i$	and	$\Gamma(A\rho_i) \cup \Gamma(B\rho_i) = E$, or
(IH2)	$A \geq_E A\rho_i$	and	$A\rho_i < B\rho_i$	and	$\Gamma(A\rho_i) \cup \Gamma(B\rho_i) = E$, or
(IH3)	$B \geq_E B\rho_i$	and	$B\rho_i < A\rho_i$	and	$\Gamma(A\rho_i) \cup \Gamma(B\rho_i) = E$, or
(IH4)	either $A\rho_i$	or	$B\rho_i$ is ground.		

Clearly, (IH1) holds for $i = 0$. Let $\{x_{i+1} \leftarrow t_{i+1}\}$ be the next mesh substituent. W.l.o.g. we assume that x_{i+1} is found in $A\rho_i$ and t_{i+1} in $B\rho_i$ in the point of disagreement between $A\rho_i$ and $B\rho_i$.

We have to investigate the following cases:

(1) (IH1) holds:

(1a) t_{i+1} is a variable:

In this case $A\rho_i \geq_E A\rho_{i+1}$ and $B\rho_i \geq_E B\rho_{i+1}$. Thus, by transitivity of the relation "$\geq_E$", $A \geq_E A\rho_{i+1}$ and $B \geq_E B\rho_{i+1}$. Moreover, the set of ground subterms of $A\rho_{i+1}$ and $B\rho_{i+1}$ remains equal to E. Therefore (IH1) also holds for i+1.

(1b) t_{i+1} is a functional term containing variables:

(The argumentation parallels that of case (1b) in the proof of lemma 5.4). By (IH1) and (*') $B\rho_i$ is weakly covering. Therefore $V(t_{i+1}) = V(B\rho_i)$. This implies that

$$B\rho_i <' A\rho_{i+1} \ (= A\rho_i \cdot \{x_{i+1} \leftarrow t_{i+1}\}).$$

Since A and B are unifiable we have $x_{i+1} \notin V(t_{i+1})$. Because $B\rho_i$ is weakly covering this implies

$$B\rho_i = B\rho_{i+1}.$$

Again, $\Gamma(A\rho_{i+1}) \cup \Gamma(B\rho_{i+1}) = \Gamma(A\rho_i) \cup \Gamma(B\rho_i) = E$.

Summarizing we have shown that, in this case, (IH3) must hold for i+1.

(1c) t_{i+1} is a ground term. There are two possibilities:

(A) Both, $A\rho_{i+1}$ and $B\rho_{i+1}$, still contain variables: In this case no new ground terms arise, i.e. $\Gamma(A\rho_{i+1}) \cup \Gamma(B\rho_{i+1}) = \Gamma(A\rho_i) \cup \Gamma(B\rho_i) = E$. Moreover, $A\rho_i \geq_E A\rho_{i+1}$ and $B\rho_i \geq_E B\rho_{i+1}$. Thus also $A \geq_E A\rho_{i+1}$ and $B \geq_E B\rho_{i+1}$. Therefore (IH1) still holds. (B) Either $A\rho_{i+1}$ or $B\rho_{i+1}$ is ground: Then clearly (IH4) holds.

(2) (IH2) holds:

(2a) t_{i+1} is a variable or a ground term:
Clearly, $A\rho_i \geq_E A\rho_{i+1}$ and $A\rho_{i+1} <' B\rho_{i+1}$.
Similiar to case (1c), either
$\Gamma(A\rho_{i+1}) \cup \Gamma(B\rho_{i+1}) = \Gamma(A\rho_i) \cup \Gamma(B\rho_i) = E$ or at least one of $A\rho_{i+1}$, $B\rho_{i+1}$ is ground. In the first case (IH2) remains valid; in the second case (IH4) holds for i+1.

(2b) t_{i+1} is a functional term containing variables:
This cannot happen because $A\rho_i <'_E B\rho_i$ implies $V(A\rho_i) \subseteq V(t_{i+1})$. This, in turn, we would mean that x_{i+1} is in $V(t_{i+1})$ which contradicts the assumption that A and B are unifiable.

(3) (IH3) holds: The argument is completely analogous to case (2).

(4) (IH4) holds: Clearly, a ground atom remains unchanged by applications of substitutions. Therefore (IH4) also holds for i+1.

We have just proved that $A \geq_E A\vartheta$ or $B \geq_E B\vartheta$, thus statements (i), (ii), and (v) of the lemma follow directly by definiton of "$\geq_E$" and τ_V, respectively. An inspection of the induction hypothesis and the single steps of the argumentation above reveals that also (iii) and (iv) hold.

Q.E.D.

We also shall make use of the following lemma:

<u>LEMMA</u> 5.6.: Let A and B be two (weakly) covering atoms s.t. V(A) = V(B). For any substitution ϑ it holds: If $A\vartheta$ is (weakly) covering then $B\vartheta$ is (weakly) covering, too.

<u>Proof:</u>
We have to show that for each functional subterm t of $B\vartheta$ either $V(t) = V(B\vartheta)$ (i.e. t contains all variables of $B\vartheta$) or, in the case of weakly covering atoms, $V(t) = \emptyset$ (i.e. t is ground).

Clearly, one of the following cases must apply:

(1) $t = s\vartheta$ for some functional subterm s of B:
If A and B are covering, then by definition, V(s) = V(B).

This implies $V(t) = V(s\vartheta) = V(B\vartheta)$.
If A and B are only weakly covering then s may also be ground although B contains variables. But in this case $t = s\vartheta$ must be ground, too.

(2) t is a subterm of some term in $rg(\vartheta)$: Since $A\vartheta$ is (weakly) covering (either t is ground or) $V(t) = V(A\vartheta)$. But, since $V(B) = V(A)$, we have $V(B\vartheta) = V(A\vartheta)$. Therefore, also in this case, $V(t) = V(B\vartheta)$.

Q. E. D.

5.2 A SIMPLE A-ORDERING STRATEGY

The concept of (weakly) covering literals permits concise definitions of some decision classes. Consider, e.g., the following class of clause sets:

5.2.1 Class E_1

DEFINITION 5.10: (E_1): A clause set S belongs to E_1 iff the following holds for all clauses C in S:

(i) All literals in C are covering, and
(ii) for all literals L, M in C either $V(L) = V(M)$ or $V(L) \cap V(M) = \emptyset$.

EXAMPLES. Let $C_1 = \{P(f(x,y), a), Q(y,x,x)\}$, $C_2 = \{P(f(x,f(x,a)), Q(z,y,a)\}$, $C_3 = \{P(x,f(a))\}$ and $C_4 = \{Q(x,y,a), P(x,x)\}$. Then $\{C_1, C_2\}$ in E_1, but any clause set containing C_3 or C_4 is not in E_1.

E_1 may be regarded as an extension of the initially extended Ackermann class (characterized by the prefix type $\exists^* \forall \exists^*$). A closely related class, called E^+, will be considered in the section below (compare also chapter 4).

We demonstrate that a simple A- ordering strategy guarantees a term depth limit for resolvents of clauses in E_1. A limit for the clause length can be achieved by splitting; thus we arrive at a resolution variant which provides a decision procedure for E_1.

The following A-ordering will prove suitable to provide a term-depth limit for covering literals.

DEFINITION S. 11.: Let A and B be two atoms, then $A <_d B$ iff

(i) $\tau(A) < \tau(B)$, and

(ii) for all x in V(A): $\tau_{max}(x, A) < \tau_{max}(x, B)$
(implying $V(A) \subseteq V(B)$).

We first have to show:

LEMMA S. 7.: $<_d$ is an A-ordering.

Proof:

Irreflexivity directly follows from the irreflexivity of "<"; transitivity holds because both "<" and "$\subseteq$" are transitive.

Let A and B be atoms s.t. $A <_d B$ and let $\sigma = \{x \leftarrow t\}$ be a component of some substitution. If $x \notin V(A)$ then $A\sigma = A$; therefore $A\sigma <_d B\sigma$ holds. If x in V(A) then, by definition of "$<_d$" $\tau_{max}(x, A) < \tau_{max}(x, B)$. Let y be any variable of V(t). We have

$\tau_{max}(y, A\sigma) = \tau_{max}(x, A) + \tau_{max}(y, t)$ and

$\tau_{max}(y, B\sigma) = \tau_{max}(x, B) + \tau_{max}(y, t)$ and therefore

$\tau_{max}(y, A\sigma) < \tau_{max}(y, B\sigma)$.

Applying some substitution ϑ to an atom has the same effect as applying all components of ϑ separately. Thus we have proved that $A <_d B$ implies $A\vartheta <_d B\vartheta$ for any ϑ.

Q.E.D.

We now show that, for every $S \in E_1$, every resolvent in $R_{<_d}(S)$ – as defined in definition S. 2 fufills the defining conditions of class E_1 and is smaller or equal w.r.t. term depth to its parent clauses. This is mainly a consequence of lemmata S. 4 and S. 6.

LEMMA 5.8.: If $E \in R_{<_d}(\{C,D\})$, where $\{C,D\}$ in E_1, then $\{E\} \in E_1$ and $\tau(E) \leq \tau(\{C,D\})$.

Proof:

Let M be the set of literals resolved upon in C and L be a literal in C - M; by ϑ we denote the m.g.u. used to generate E. By definition any two literals of C share either all or none of their variables. We have to consider the following cases:

(1) $V(L) \cap V(M) = \emptyset$: In this case $\vartheta(x) = x$ for all x in $V(C - M)$ which implies $L\vartheta = L$. Thus L occurs unchanged in E.

(2) $V(L) = V(L')$ for some $L' \in M$:

(2a) $L <_d L'$: As "$<_d$" is an A- ordering we have $L\vartheta <_d L'\vartheta$ which implies $\tau(L\vartheta) < \tau(L'\vartheta)$. Thus by lemma 5.4 also $\tau(L\vartheta) < \tau(\{C,D\})$. Moreover it follows from lemma 5.6 that $L\vartheta$ is covering.

(2b) $L \not<_d L'$: By definiton of $R_{<_d}$ we also know that $L' \not<_d L$. Now observe that in a covering atom A all variables occur somewhere in maximal depth; i.e. for all x in $V(A)$: $\tau_{max}(x, A) = \tau(A)$. This implies $\tau(L) = \tau(L')$ and $\tau(L\sigma) = \tau(L'\sigma)$ for any substitution σ. Moreover lemmata 5.4 and 5.6 again guarantee that $L\vartheta$ is covering.

By analogy the same holds for the literals in D. Thus we have proved that E fullfills the relevant conditions.

Q.E.D.

If we use $R_{<_d}$ in combination with the splitting rule (c. f. definition 2.29) we arrive at the following theorem:

THEOREM 5.9.: Class E_1 is decidable; $R_{<_d}$ combined with the splitting rule provides a decision procedure.

Proof:

Lemma 5.8 provides a term depth limit and guarantees that the resolvents are in E_1 again. What is still missing in order to obtain the decidability of E_1 is a length limit for the resolvents. To this aim we employ the splitting rule.

In order to decide whether a clause set $S \in R_{<_d}$ is satisfiable we first construct SPLIT(S) and then apply $R_{<_d}$ to each S' in SPLIT(S).

Because of condition (ii) in the definition of E_1 the set of variables V(L) is the same for all literals L of a split component of a clause. This fact together with part (iii) of lemma 5.4 implies

$$|V(E)| \leq \max(|V(C)|, |V(D)|)$$

for resolvents E of clauses C and D that are split components of clauses in class E_1. Considering the term depth limit for the resolvents as expressed by lemma 5.8 we arrive at a length limit because, under equivalence w.r.t. renaming of variables, there are only finitely many literals L if $|V(L)|$ as well as $\tau(L)$ are bounded. (Remember that we only consider finite clause sets and thus may assume that there are only finitely many predicate letters and function symbols).

Q.E.D.

5.3 CLASS E^+

The following class was investigated first by T.Tammet in [Tam90] (compare also chapter 4):

DEFINITION 5.12.: (E^+): A clause set S belongs to E^+ iff the following holds for all clauses C in S:

(i) All literals in C are weakly covering, and

(ii) for all literals L, M in C either $V(L) = V(M)$ or $V(L) \cap V(M) = \emptyset$.

EXAMPLES. Let $C_1 = \{P(f(x,y), a), Q(y,x,x)\}$, $C_2 = \{P(f(x,f(x,a)), Q(z,y,a)\}$, $C_3 = \{P(x,f(a))\}$ and $C_4 = \{Q(x,y,a), P(x,x)\}$. Then $\{C_1, C_2, C_3\}$ in E^+, but any clause set containing C_4 is not in E^+.

E^+ clearly is a superset of E_1, because it additionally allows arbitrary ground terms to occur everywhere in the clauses. E^+ also contains the class of clause sets only consisting of clauses C s.t. $|V(C)| \leq 1$. This subclass seems to have been proved decidable first by Y. Gurevich [Gur73] and is often called class E, which motivated the name E^+ for the class defined above.

T. Tammet [Tam90] showed that a special ordering strategy terminates on all sets in E^+ (i.e. there are only finitely many resolvents). But, as already mentioned in chapter 4, Tammet's refinement is neither an A- ordering nor even a π - ordering strategy and no completeness proof for this refinement could be achieved so far. Thus the decidability of E^+ remained an open problem until now. Using a semantic tree argument we shall show that a simple strengthening of the A-ordering strategy of the last section combined with a saturation rule is both, terminating and complete, on clause sets of E^+. (To mention an open problem we conjecture that already the $R_{<_d}$ -refinement itself is sufficient to guarantee the terminiation of the resolution procedure; however this seems hard to prove). We start by defining explicitly the ordering that arises if condition (i) of the definition of $<_d$ (definition 5.11) is dropped for non ground atoms:

<u>DEFINITION</u> 5.13.: For any two atoms A and B $A <_{vd} B$ iff

(i) $\tau(A) < \tau(B)$ whenever A and B are ground, and

(ii) for all x in V(A): $\tau_{max}(x, A) < \tau_{max}(x, B)$
(implying $V(A) \subseteq V(B)$).

<u>Remark</u>: Condition (i) is neither necessary for the termination proof nor for the completeness proof below. But since it makes the ordering condition more restrictive and thus results in a more efficient resolution procedure we prefered to include it. Observe that for non ground, weakly covering atoms A and B: $A <_{vd} B$ iff $\tau_v(A) < \tau_v(B)$.

We define $R_{<_{vd}}$ analogously to the A- ordering refinements:

<u>DEFINITION</u> 5.14.: For any clause set S: E is in $R_{<_d}(S)$ iff E is a Robinson-resolvent of clauses in S s.t. for no atom B in E: $A <_{vd} B$, where A is the resolved atom.

Unlike in the case of E_1, the $R_{<_{vd}}$ - resolvents of clauses of E^+ may be deeper w.r.t. term depth than their parent clauses.

<u>EXAMPLE</u>: Let $C = \{\neg P(x, g(x, y)), P(g(x, y), f(a))\}$ and $D = \{P(f(a), g(u, v))\}$. Then $E = \{P(g(f(a), y), f(a))\}$ is an $R_{<_{vd}}$ - resolvent of C and D; but $\tau(E) = 2$ whereas $\tau(C) = \tau(D) = 1$.

However, we shall show that, for $R_{<_{vd}}$ -resolvents, the maximum depth of occurrences of variables (τ_v) cannot increase (compared to the corresponding parent clauses) and that this fact suffices to guarantee the termination of our resolution procedure.
We remark that the investigation of E^+ as carried out in chapter 4 is quite similar to the one presented here.

<u>THEOREM</u> 5.10.: $R_{<_{vd}}$ terminates if applied to decomposed clause sets S in E^+, i.e. $R^*_{<_{vd}}(S)$ is finite for all S in E^+.

<u>Proof</u>:
We start by proving that for any decomposed S in E^+ the set $R_{<_{vd}}(S)$ is in E^+, again. Let E be some clause of $R_{<_{vd}}(S)$ and A be the resolved atom that was used to form E. By the definition of E^+ and part (i) of lemma 5.5, A is weakly covering. Therefore all literals in E are weakly covering by lemma 5.6. The fact that $V(A) = V(B)$ implies $V(A\sigma) = V(B\sigma)$ for any atoms A, B and any substitution σ guarantees that {E} in E^+. Clearly, if E is not ground it is again decomposed. Otherwise, by the definition of $R_{<_{vd}}$ and E^+, only new ground clauses, not deeper than their parent clauses, may be derived using E. Since there are only finitely many such clauses we may, w.l.o.g., in the following assume that the clauses are not ground.

Observe that, by part (ii) of lemma 5.5, $\tau_{v(A)} \leq \tau_v(\{C,D\})$. By the ordering condition of the definition of $R_{<_{vd}}$ this implies that $\tau_v(E) \leq \tau_v(\{C,D\})$. We have thus proved that $\tau_v(R^*_{<_{vd}}(S)) \leq \tau_v(S)$. This means that there is a (global) limit for the maximum depth of variable occurrences in $R_{<_{vd}}$ - resolvents of any clause set in E^+.

Now observe that there also is only a finite number of different ground terms that may occur in $R_{<_{vd}}$ - resolvents: Indeed, it follows from part (iv) of lemma 5.5 that (non ground) resolvents contain only ground terms that already occur somewhere in the respective parent clauses.

Combining these two results we arrive at a (global) term depth limit for the resolvents.

Since, by part (v) of lemma 5.5, the number of variables in resolvents of decomposed clauses can not surpass that of its parent clauses we have shown that — up to renaming of variables - only finitely many resolvents can be derived by $R_{<vd}$.

Although we do not know whether $R_{<vd}(S)$ is a complete resolution refinement — at least if restricted to E^+ — we shall show below that the combination of a special saturation rule and $R_{<vd}(S)$ is complete on E^+.

DEFINITION 5.15.: Let L_1 and L_2 be weakly covering literals s.t. $V(L_1) = V(L_2)$. (L_1, L_2) is called a critical pair iff $\tau_v(L_1) < \tau_v(L_2)$ and $\tau(L_1) \geq \tau(L_2)$.

To motivate this definition and that of our saturation rule below, observe that for all clauses C and D fulfilling the defining conditions of E^+ and any $R_{<d}$-resolvent E of C and D: $\tau_v(E) \leq \tau_v(\{C,D\})$ unless either C or D contains a critical pair of literals. But, in contrast to $<_d$, we have $A <_{vd} B$ whenever (A, B) is a critical pair. It is this fact that prevents the generation of an $R_{<vd}$ - resolvent E of C and D s.t. $\tau_v(E) > \tau_v(\{C,D\})$.

EXAMPLES: Let $L_1 = P(x, y, f(f(a)))$, $L_2 = Q(x, g(x,y), f(f(a)))$, and $L_3 = P(f(f(f(a))), x, y)$. Then (L_1, L_2) and (L_3, L_2) are critical pairs, but no other pair of literals in $\{L_1, L_2, L_3\}$ is critical.

We define a saturation operator that will help us to turn $R_{<vd}$ into a provable complete resolution variant:

DEFINITION 5.16.: For any clause C in some clause set S of E^+ we define:

(i) If C contains a critical pair (L_1, L_2) then $FILL_S(C) = \{C\sigma \mid \sigma$ is a ground substitution, based on S (c.f definition 2.17) s.t. $\tau(L_1\sigma) \geq \tau(L_2\sigma)$ for all critical pairs $(L_1, L_2)\}$.

(ii) Otherwise $FILL_S(C) = \emptyset$.

For a clause set S we define $FILL(S) = S \cup \bigcup_{C \in S} FILL_S(C)$.

EXAMPLES: Let $L_1 = P(x, f(f(a)))$, $L_2 = Q(x, f(x), f(f(a)))$, and $L_3 = P(f(f(f(a))), x)$. Let $C = \{L_1, L_2, L_3\}$ be a clause in S, where $CS(S) = \{a\}$ and $FS(S) = \{f\}$. Then

$$FILL_S(C) = \{\{P(a, f(f(a))),\ Q(a, f(a), f(f(a))),\ P(f(f(f(a))), a)\},$$
$$\{P(f(a), f(f(a))),\ Q(f(a), f(f(a)), f(f(a)))\ \ P(f(f(f(a))), f(a))\},$$
$$\{P(f(f(a)), f(f(a))),\ Q(f(f(a)), f(f(f(a))), f(f(a))),\ P(f(f(f(a))), f(f(a)))\}\}.$$

Observe that FILL(S) is always finite for finite clause sets S.

We define a resolution variant that combines $R_{<vd}$ with the fill operator:

DEFINITION 5.17.: For any clause set S:

$$R_{FILL}(S) = FILL(R_{<vd}(S)).$$

The members of $R_{FILL}(S)$ are called R_{FILL}-resolvents of S.

THEOREM 5.11.: Resolution procedure R_{FILL}, combined with the splitting rule, is complete for all clause sets S in E^+.

Proof:

Considering the remarks in section 5.1.2 about completeness via semantic trees (see especially lemma 5.2) we know that it is sufficient to prove the following:

(*) For any unsatisfiable clause set S there is a semantic tree T based on some $K \subseteq \hat{H}_S$ and closed for S s.t. any two clauses which fail immediately below an inference node of T have an R_{FILL}-resolvent which fails at that inference node.

Let T be a semantic tree closed for S, based on a subset K of $\hat{H}_S$ s.t. the atoms in any path from the root down to some other node of T are ordered according to their term depth. With other words, whenever $\tau(A) < \tau(B)$ for A, B in K then the edges of T labeled with A are situated above, i.e. closer to the root than, the edges labeled with B.

Let C and D be clauses in S failing immediately below, but not at, a node n of T. Let γ_C and γ_D be the two ground substitutions s.t. $C\gamma_C$ and $D\gamma_D$, respectively, are subsets of the refutation sets of the corresponding nodes. Finally, let A_k and $\neg A_k$ be the two ground literals labeling the edges leaving the inference node n. (W.l.o.g. we may assume that $A_k \in C\gamma_C$ and $\neg A_k \in D\gamma_D$). By theorem 5.2 we know that there exists a Robinson-resolvent E of C and D that fails at n. Let A' be the corresponding resolved atom. We have to show that there also exists an R_{FILL} - resolvent E' failing at n.

Consider the following cases:

(1) Neither in C nor in D there is a critical pair (L_1, L_2) s.t. L_1 is among the literals resolved upon: In this case $\tau(A') < \tau(B)$ iff $A' <_{vd} B$ for all atoms B in E. By the definiton of the semantic tree, A_k is of maximal term depth within the literals of the refutation sets of the sons of n. A' is an instance of A_k Therefore, by the defining conditions for clauses in sets of E^+, also $\tau(A) \geq \tau(E)$. This implies that E itself is the R_{FILL} - resolvent we are looking for.

(2) There is a critical pair (L_1, L_2) in C, s.t. L_1 is one of the literals resolved upon in C in the derivation of E:

(2a) $\tau(L_1\gamma_C) \geq \tau(L_2\gamma_C)$: This means that $C\gamma_C \in FILL_S(C)$. In this case, consider the $R_{<d}$-resolvents of $C\gamma_C$ and D: They are all in $R_{FILL}(S)$ and, by definition of the semantic tree, at least one of them fails at n.

(2b) $\tau(L_1\gamma_C) < \tau(L_2\gamma_C)$: We show that this cannot happen. Because we assumed that L_1 is one of the literals resolved upon in C we have $L_1\gamma_C = A_k$. The semantic tree is defined s.t. $\tau(A_k) \geq \tau(\{C\gamma_C, D\gamma_D\})$. But this contradicts $\tau(L_1\gamma_C) < \tau(L_2\gamma_C)$.

Q.E.D.

Given theorems 5.10 and 5.11 it is an easy task to prove the following:

<u>THEOREM</u> 5.12.: E^+ is decidable; R_{FILL} combined with the splitting rule provides a decision procedure.

<u>Proof</u>:

The only thing we need to observe is that -- on clause sets in E^+-- R_{FILL} behaves exactly like $R_{<vd}$ (S) except for the additional generation of ground clauses that are always limited in term depth by the depth of their parent clauses. Thus we conclude from theorem 5.10 that also $R^*_{FILL}(S)$ is finite for all decomposed clause sets S in E^+. Now it follows directly from theorem 5.11 that $R_{<vd}$ yields a decision procedure.

Q.E.D.

5.4 A-ORDERINGS COMBINED WITH SATURATION RULES

5.4.1 An extension of the Skolem class

The initially extended Skolem class is the class of prenex formulas with a prefix of the form $\exists z_1 \ldots \exists z_l \forall y_1 \ldots \forall y_m \exists x_1 \ldots \exists x_n$ s.t. each atom of the matrix has among its arguments either

(i) at least one of the x_i, or

(ii) at most one of the y_i, or

(iii) all of $y_1, \ldots, y_m$.

In this section we consider a class which strongly generalizes the initially extended Skolem class. Consider the following definition:

<u>DEFINITION</u> 5.18.: ($\mathcal{S}^+$) A clause set S belongs to $\mathcal{S}^+$ iff for all clauses C in S and all literals L in C:

(i) If t is a functional term occurring in C then $V(t) = V(C)$, and

(ii) $|V(L)| \leq 1$ or $V(L) = V(C)$.

Observe that condition (i) is equivalent to:

(i') If L is functional, then L is covering and $V(L) = V(C)$.

Class $\mathcal{S}^+$ not only contains the initially extended Skolem class but also the initially extended Gödel class (i.e. the prefix class with quantifier prefix type $\exists^* \forall \forall \exists^*$).

$\mathcal{S}^+$ is related to E_1. In fact we have (analogously to lemma 5.8 in section 5.2.1:

$$E \in R_{<_d}(\{C,D\}) \text{ implies } \tau(E) \leq \tau(\{C,D\})$$

if $\{C,D\} \in \mathcal{S}^+$. The only problem is that in general $R_{<_d}(S) \notin \mathcal{S}^+$ for clause sets $S \in \mathcal{S}^+$. Atoms may be generated which, besides covering terms, have arbitrary variables as arguments.

To be able to argue more accurately we define:

DEFINITION 5.19.: An atom or literal A is called essentially monadic on a term t iff t is an argument of A and each other argument is either equal to t or a constant. A is called almost monadic on t iff t is functional and - besides t and constants - also some variables that are not in V(t) occur among the arguments of A. More precisely, A is almost monadic on some functional term t iff $t \in args(A)$, $\forall_{s \neq t \in args(a)}$: s is a constant or a variable $\notin V(t)$.

EXAMPLES: $P(g(x), b, g(x))$ is essentially monadic on $g(x)$. $P(f(f(z)), x, a, f(f(z)))$ is almost monadic on $f(f(z))$. $P(x, y, a, x)$ and $Q(f(x), f(y))$ are neither essentially monadic nor almost monadic (on some subterm).

As mentioned above $R_{<d}$ provides a term depth limit for the resolvents of $\mathcal{S}^+$. But by resolving such clauses, almost monadic atoms may be generated besides covering ones.

EXAMPLE: Let $C = \{P(x), Q(x, y)\}$ and $D = \{\neg P(f(z)), Q(g(z), z)\}$. The only $R_{<d}$-resolvent of C and D is $E = \{Q(f(z), y), Q(g(z), z)\}$. The first literal of E is almost monadic on $f(z)$; the second literal is covering and essentially monadic.

It is interesting to mention that (for class $\mathcal{S}^+$) almost covering atoms are generated by $R_{<d}$ only if one of the parent clauses is function free, and the atom(s) resolved upon in this clause contain(s) one variable only, whereas other atoms must have also additional variables as arguments. In all other cases all atoms of a resolvent are covering. It would be an easy task to refine the ordering $<_d$ in a way such that it becomes sufficiently restrictive to decide $\mathcal{S}^+$ (and many other classes). We would just have to add

(iv) $V(A) \subset V(B)$ implies $A <_d B$ for function free atoms A and B

to the defining conditions for $<_d$ (definition 5.11). Unfortunately the resulting resolution variant is neither an A-ordering nor an π - ordering strategy.

Special orderings, similiar to this, are treated in chapter 4. Although no ground projection of such orderings exists in general, the resulting strategies may nevertheless be complete on certain classes of clause sets. But this cannot be demonstrated in the realm of the semantic trees. Since, in this chapter, we are interested in strategies that are complete via semantic trees we have to go a different way: We shall combine $R_{<_d}$ with a special saturation rule. For any almost monadic atom we define a corresponding set of essentially monadic atoms.

DEFINITION 5.20.: Let A be almost monadic on some functional term t and Con be some set of constants. The monadization MON(A, Con) consists of atoms that are like A except for replacing each variable that occurs as argument of A (but not of t) by t or some constant in Con. More exactly:

Let $\Sigma_{t,Con}$ be the set of all substitutions of the form

$\{t_i/x_i \mid x_i \in V(A) - V(t)\}$ where $t_i = t$ or $t_i \in Con$ then

$MON(A,Const) = \{A\sigma \mid \sigma \in \Sigma_{t,Con}\}$.

We extend the definition of MON to clauses and clause sets: If all almost monadic atoms A of a clause C are almost monadic on the same functional term t then $MON(C, Con) = \{C\sigma' \mid \sigma' \text{ in } \Sigma'_{t,Con}\}$ where $\Sigma'_{t,con}$ is the set of substitutions $\{t_i/x_i \mid x_i \in V(C) - V(t)\}$, s.t. $t_i = t$ or $t_i \in Con$.

If C is function free and there is one and only one variable x, s.t. all literals $L \in C$ with $|V(L)| \geq 2$ contain x then $MON(C, Con) = \{C\sigma \mid \sigma \text{ in } \Sigma_{x,Con}\}$ where $\Sigma_{x,Con}$ is the set of substitutions $\{t_i/x_i \mid x_i \in V(C) - V(t)\}$, s.t. $t_i = x$ or $t_i \in Con$. In all other cases $MON(C, Con) = \{C\}$.

For any clause set S: $MON(S) = \bigcup_{C \in S} MON(C, CS(S))$
(where CS(S) is the set of all constants occurring in clauses of S).

Remark: As we only use the monadisation operator in the context of finite clause sets S we will implicitly assume that the set of constants occurring in S is always used for the monadisation of atoms or clauses. For sake of readability we therefore suppress the second argument and write MON(A) and MON(C) in stead of MON(A, Con) and MON(C, Con) , respectively. Observe that, in our context, MON(A) and MON(C) are always finite.

<u>EXAMPLES</u>: For all examples we assume that there is just one constant a.
Let $A = P(f(x,y), z)$ then $MON(A) = \{P(f(x,y), f(x,y)), P(f(x,y), a)\}$.
Let $C = \{Q(u,x,f(x,a)), P(u, f(x,a))\}$ then
$MON(C) = \{\{Q(f(x,a), x, f(x,a)), P(f(x,a), f(x,a))\}, \{Q(a,x,f(x,a)), P(a,f(x,a))\}\}$.
For $D = \{P(x,x), Q(u,a,x), P(x,v)\}$ we have $MON(D) = \{\{P(x,x), Q(x,a,x)\}$, $\{P(x,x), Q(x,a,x), P(x,a)\}, \{P(x,x), Q(a,a,x)\}, \{P(x,x), Q(a,a,x), P(x,a)\}\}$.

We may now define a new resolution variant R_m which is based on $R_{<_d}$.

<u>DEFINITION</u> 5.21: For any clause set S:

$$R_m(S) = MON(R_{<_d}(S)).$$

The members of $R_m(S)$ are called R_m- resolvents.

We state:

<u>LEMMA</u> 5.13.: If $\{C,D\} \in \mathcal{S}^+$ and $E \in R_m(\{C,D\})$ then $\tau(E) \leq \tau(\{C,D\})$.

<u>Proof</u>:
As mentioned above, each atom of an $R_{<_d}$-resolvent of any clause set in $\mathcal{S}^+$ is either almost monadic or covering. There are two cases:

(1) If E is an $R_{<_d}$ - resolvent of {C,D} and does not contain almost monadic atoms: Then E is an R_m- resolvent, too, and $\tau(E) = \tau(\{C,D\})$ follows directly from lemma 5.4.

(2) Let E be an $R_{<_d}$-resolvent generated using the resolved atom A and let E contain some almost monadic atom B. In this case one of the parent clauses must be function free and the literals resolved upon in that clause may only contain one variable. Moreover B is the instance of an atom in that clause, containing additional variables. It follows that A is essentially monadic on some term t and B is almost monadic on t. Therefore $\tau(B) = \tau(A)$. By definition we have $\tau(B') = \tau(B)$ for all $B' \in MON(B)$. (The same holds for any almost monadic atom of E). By lemma 5.4 $\tau(A) \leq \tau(\{C,D\})$. Therefore we conclude that $\tau(E') \leq \tau(\{C,D\})$ for all $E' \in MON(E)$.

Q.E.D.

In order to prove that the resolvents of clause sets in $\mathcal{S}^+$ are in $\mathcal{S}^+$ again, we make use of the well known splitting mechanism (see section 2.6). We want to remark that even without splitting R_m decides class $\mathcal{S}^+$. We employ the splitting rule to make the proof somewhat simpler.

LEMMA 5.14.: If S in $\mathcal{S}^+$, then all members of SPLIT(R_m(S)) are in class $\mathcal{S}^+$, too.

Proof:

Let C, D in S and let A be an atom resolved upon to generate some resolvent E in $R_{<_d}\{C,D\}$. Let ϑ be the m.g.u. used to get E. (Thus $A\vartheta$ is the resolved atom). We have to consider the following cases:

(1) $A\vartheta$ is function free: By definition of $R_{<_d}$ and $\mathcal{S}^+$, this only occurs if both, C and D, are function free. Therefore E is function free, too.

(1a) $|V(A\vartheta)| = 0$:

In this case $V(C\vartheta) \cap V(D\vartheta) = \emptyset$. But this implies that for any two atoms B_1, B_2 in E s.t. $|V(B_i)| > 1$ but $V(B_i) \neq V(E)$ $(i = 1, 2)$ either $V(B_1) \cap V(B_2) = \emptyset$ or, by definition of $\mathcal{S}^+$, $V(B_1) = V(B_2)$. Thus the split components of E fulfill the defining conditions of class $\mathcal{S}^+$.

(1b) $|V(A\vartheta)| = 1$:

Let x be the only element of $V(A\vartheta)$. Then for any literal $L \in E$ s.t. $|V(L)| \geq 2$ we have $x \in V(L)$. Thus, by definition of monadisation, $|V(E')| \leq 1$ for all $E' \in MON(E)$. Moreover, all clauses in MON(E) are function-free, too. This implies that the R_m-resolvents and accordingly also their split components are in class $\mathcal{S}^+$.

(1c) $|V(A\vartheta)| > 1$:

In this case E itself is an R_m - resolvent and fulfills the relevant conditions.

(2) $\tau(A\vartheta) > 0$:

W.l.o.g. we may assume that A is in C and just investigate what happens with literals of C when ϑ is applied.

(2a) $|V(A)| = 1$: If all literals of C contain at most one variable the defining conditions of class $\mathcal{S}^+$ clearly remain satisfied after applying ϑ and splitting. The only interesting case arises when there is some $B \in C$, s.t. $|V(B)| > 1$. As A only contains one variable and – by definition of $\mathcal{S}^+$ – no functional ground term, it follows that $A\vartheta$ is essentially monadic on some functional term t. Therefore $B\vartheta$ is almost monadic on t. By the definition of the monadisation operator $V(B') = V(t) = V(A\vartheta)$ for all $B' \in MON(B\vartheta)$. This holds for all such literals. Therefore $V(A\vartheta) = V(E')$ for all $E' \in MON(E)$, which was to show.

(2b) $|V(A)| > 1$:

It follows from the definition of $\mathcal{S}^+$ that A is covering and that $V(A) = V(C)$. For each $B \in C$ we have either

(i) $V(B) = V(A)$, which implies that $V(B\vartheta) = V(A\vartheta)$, or

(ii) B is not functional and contains just one variable. Call this variable x.

By the proof of lemma 5.6 $\vartheta(x)$ is either a variable, a constant or a functional term t, s.t. $V(t) = V(A\vartheta)$.

Since $A\vartheta$ contains all variables that occur in any functional term of $C\vartheta$ or $D\vartheta$ we have $V(A\vartheta) = V(E')$ for all $E' \in MON(E)$. Therefore the defining conditions of class $\mathcal{S}^+$ remain satisfied for the respective resolvents.

Q.E.D.

We may now state that R_m decides $\mathcal{S}^+$:

<u>THEOREM</u> 5.15.: $\mathcal{S}^+$ is decidable; R_m combined with the splitting rule provides a decision procedure.

<u>Proof</u>:

Lemma 5.13 guarantees a term depth limit for R_m-resolvents of clause sets in $\mathcal{S}^+$. Lemma 5.14 shows that the resolvents (at least after splitting) are in $\mathcal{S}^+$ again. It remains to establish a length limit for the resolvents: By lemma 5.4 and the definition of $\mathcal{S}^+$ it follows that the number of variables of an $R_{<d}$ -resolvent is bounded by the number of variables of its parent clauses. Clearly, this number can never increase through splitting or monadisation. Taking into account the term depth limit, this guarantees a length limit for the R_m - resolvents. This in turn implies that $R_m^*(S)$ is finite for any finite clause set S. Thus the theorem follows from the completeness of R_m on clause sets of $\mathcal{S}^+$ (cf. section 5.4.2 below).

Q.E.D.

5.4.2 Completeness of R_m

To justify the theorem 5.15 it remains to prove the following.

THEOREM 5.16.: The resolution procedure R_m, combined with the splitting rule, is complete for all clause sets S in $\mathcal{S}^+$.

Proof:
By lemma 5.2 our task is to show:

(1) For any unsatisfiable clause set S there is a semantic tree T based on some $K \subseteq \hat{H}_S$ and closed for S s.t. any two clauses which fail immediately below an inference node of T form an R_m-resolvent which fails at the inference node.

For any clause set S in $\mathcal{S}^+$ we define enumerations $A_0, A_1, \ldots$ of arbitrary subsets of the corresponding Herbrand base s.t. less deep atoms precede deeper ones. Within atoms of equal depth essentially monadic atoms precede those atoms that are not essentially monadic.

More formally we have:

(i) if $\tau(A_i) < \tau(A_j)$ then $i < j$, and
(ii) if $\tau(A_i) = \tau(A_j)$ and $i < j$ then A_i is essentially monadic whenever A_j is essentially monadic.

Note that "$<_d$" is compatible with all such enumerations. Let $S \in \mathcal{S}^+$ be unsatisfiable and let T be a semantic tree for some subset K of $\hat{H}_S$ closed for S s.t. T is based on such an enumeration. (Theorem 5.1 guarantees the existence of T). Let C and D be clauses in S failing immediately below, but not at, a node n of T. Let A_k and $\neg A_k$ be the two ground literals labeling the edges leaving the inference node n. Because "$<_d$" is an A-ordering there exists an $R_{<_d}$-resolvent E of C and D, s.t. A_k is a ground instance of the resolved atom A, which fails at n. By definition of $\mathcal{S}^+$ and lemma 5.4 each atom of E either is covering or almost monadic on some term. If for all atoms B of E, B is covering and either $|V(B)| \leq 1$ or $V(B) = V(E)$ then E is an R_m-resolvent, too, and the theorem clearly holds.

There are two crucial cases:

(1) The $R_{<_d}$ - resolvent E contains literals that are almost monadic on some term: This may only be the case if A is essentially monadic on some term t, and all atoms of E are either almost monadic on t or covering. We have to show that some clause in MON(E), fails at node n.

Let γ be the substitution s.t. $E\gamma$ is in the refutation set of n. Clearly $A\gamma$ is essentially monadic on $t\gamma$. For any almost monadic atom B of E, $B\gamma$ is of the same depth as $A\gamma$ because all functional arguments of A as well as of B are equal to t. Therefore, by definition of the enumeration of the Herbrand base, also $B\gamma$ is essentially monadic. This implies that $B'\gamma = B\gamma$ for some $B' \in MON(B)$. Thus some $E' \in MON(E)$ fails at node n, too.

(2) E is function free and for some literal L in E we have $|V(L)| > 1$ but $V(L) \neq V(E)$ (i.e. condition (ii) of the definition of class $\mathcal{S}^+$ does not hold). There are two subcases:

(2a) The resolved atom A is ground: In this case we can split E into components that fulfill the relevant conditions.

(2b) A contains just one variable x:
Then all atoms of E with more than one variable contain x, too. By definition of MON all variables of E are replaced by x or by constants. As A is function free and $|V(A)| = 1$ we know that $A\gamma$ is essentially monadic on some term t. Clearly $\gamma(x) = t$. Therefore, by the construction of T, also $L\gamma$ is essentially monadic on t. Observe that, by definition of $\mathcal{S}^+$, for every literal $M \in E$, s.t. $|V(M)| = 1$ there is some literal L, s.t. $V(M) \subset V(L)$ and $x \in V(L)$. Therefore all literals of $E\gamma$ are essentially monadic on t or some constant. It follows, like in case (1), that $E'\gamma = E\gamma$ for some $E' \in MON(E)$.

Q.E.D.

5.5 PURE SATURATION STRATEGIES

5.5.1 The Bernays – Schönfinkel class

As already mentioned in section 3.2 the Bernays-Schönfinkel class is the class of prenex formulas with quantifier prefixes of type $\exists^* \forall^*$ (cf. [BS28]). This corresponds to the class of function free clause sets, which we shall call BS* in the following. Since the Herbrand universe is always finite for such clause sets the decidability of BS* is a trivial consequence of Herbrand's theorem. Nevertheless there seems to be no direct possibility to use a resolution strategy as a decision procedure. We have already seen in section 3.2 that no semantic resolution refinement (in the sense of chapter 3) can decide BS*.

The following example will make clear that also A-ordering refinements combined with the usual deletion strategy cannot always limit the length of the resolvents on class BS*.

<u>EXAMPLE</u>: Let $E_2 = \{P(x_0, x_1), P(x_1, x_2), \neg P(x_2, x_0)\}$ and $D = \{P(y_2, y_0), P(y_2, y_3), \neg P(y_3, y_0)\}$. For all $n > 2$:

$$E_n = \{P(x_0, x_1), P(x_1, x_2), \ldots, P(x_{n-1}, x_n), \neg P(x_n, x_0)\}$$

is a resolvent of E_{n_1} and D. Observe that no atom ordering restriction that has a ground projection can prevent the generation of this sequence of resolvents. Moreover, for no $i \neq j$ E_i subsumes E_j nor is any clause E_i a tautology. This demonstrates that we cannot use A-ordering strategies to decide BS.

For sake of completeness we mention that for any S we might of course generate explicitly the finite set S_g of all Herbrand instances of the clauses in S and afterwards apply resolution to this set of ground clauses to check whether S is satisfiable or not. Replacing clauses by certain instances of it, is what we call a saturation rule. But if there are many different constants and some predicate symbols of higher arity in S, this saturation strategy clearly is impracticable. We just mention that more feasible strategies could be achieved by combining semantic resolution or A-ordering refinements with (partial) saturation.

Such a strategy is presented in chapter 3.2, where BS* is translated to PVD (via partial saturation) and afterwards decided by semantic clash resolution.

5.5.2 Uniform atoms and clauses

To support a concise formulation of the decision class to be stated in the next subsection we introduce some additional terminology:

DEFINITION 5.22.: Two terms t_1, t_2 are called congruent iff both terms are functional and the sequence of arguments of t_1 is a permutation of the sequence of arguments of t_2.

DEFINITION 5.23.: An atom or literal A is called uniform iff

(i) A is function free, or

(ii) there is a functional argument t of A, s.t. each argument of A is either congruent to t, an argument of t or a constant.

EXAMPLES. $P(a, x, f(x,y), g(y,x))$ and $P(u, u, a, v)$ are uniform literals, but $Q(z,f(x,y))$ is not uniform.

Remark. The definition of uniform literals generalizes the concept of argumental literals as introduced by Joyner [Joy76]: It is easy to show that all argumental literals are uniform.

DEFININITION 5.24.: A substitution σ s.t. $rg(\sigma)$ consists of variables and constants only is called flat.

We state some simple properties of uniform atoms:

LEMMA 5.17.:

(1) If ϑ is an m.g.u. of two uniform atoms A, B then $A\vartheta$ (= $B\vartheta$) is uniform, too.

(1) If ϑ is an m.g.u. of two uniform and functional atoms A, B then ϑ is flat.

Proof.

Follows immediately from definition 5.23.

Resolving two uniform clauses does not always result in a uniform clause again. The following example will help to unterstand the situation.

EXAMPLE. Let C = { P(x,y), Q(x,z), Q(z,w) } and D = { ¬P(f(u,v),v), P(g(v,u),u)}. C and D are uniform but their resolvent E = {Q(f(u,v),z), Q(z,w), P(g(v,u),u)} is not uniform, because there are variables among the arguments of the literals that do not occur in the functional terms f(u,v) and g(u,v), respectively.

5.5.3 The classes U and M^+

Consider the following class:

DEFININITION 5.25.: A clause set S belongs to U iff for all clauses C in S:

(i) Each L in C is uniform,

(ii) $\tau(C) \leq 1$ (i.e. there is no nesting of function symbols).

Clearly U is an undecidable class: It, e.g., contains the clause form of the Skolem class, i.e., the class of prenex formulas (without function symbols) with prefix of type $\exists^*\forall^*\exists^*$. However we shall demonstrate the following subclass of U to be decidable.

DEFININITION 5.26.: A clause set S belongs to M^+ iff S in U and all clauses C in S are Krom, i.e. $|C| \leq 2$.

M^+ is an extension of the Maslov class, i.e. the class of prenex formulas that are Krom and have a Skolem-type prefix (i.e., $\exists^*\forall^*\exists^*$). Obviously, we arrive at clause sets in M^+ if we Skolemize formulas of the Maslov class. Maslov formulas and hence also clause sets in M^+ can be used to encode quite naturally computations for different machine models (see e.g. [DL83]).

To decide M^+ we define a kind of resolution variant R_U which replaces each ordinary resolvent by all of its instances that fulfill the defining conditions of U.

DEFININITION 5.27.: For any clause set S the set of R_U- resolvents of S, $R_U(S)$, is the set of all clauses $E\sigma$, where E is an R−resolvent of clauses in S and σ is a substitution based on S s.t. $\{E\sigma\} \in$ U.

Resolution strategies R' are usually defined such that, for any clause set S, the set of resolvents R'(S) is a proper subset of R(S) (i.e. the set of Robinson resolvents). In contrast, $R_U(S)$ contains certain instances of R-resolvents. As any clause subsumes each of its instances this provides a correct deduction mechanism. But it is quite obvious that R_U is not complete for arbitrary clause sets; i.e. the empty clause may not be derivable from S although S is unsatisfiable. However we shall show that R_U is complete as long as we consider clause sets in U only. This has the interesting consequence that although U is an undecidable class we only have to take into account resolvents of a limited term complexity (depth 1). Of course this implies that there is no recursive bound on the length of resolvents needed to refute a clause set in U using R_U.

Observe that each clause set of M^+ is finite - modulo renaming of variables - as long as the number of different constants, predicate and function symbols is finite. Therefore, by iteratively generating all R_U - resolvents starting with a clause set S in M^+, we arrive at a finite set $R_U^*(S)$ of clauses s.t. no new R_U -resolvents can be derived from $R_U^*(S)$. If $R_U^*(S)$ contains the empty clause then S is unsatisfiable, otherwise S is satisfiable, provided that R_U is complete for clause set S. In other words, the following theorem is a corollary to theorem 5.21 which we shall prove in section 5.5.4 below:

<u>THEOREM</u> 5.18.: R_U - resolution provides a decision procedure for M^+.

From a practical point of view the resolution variant R_U, although definitely simpler than Joyner's correponding R_3 (c.f. [Joy76]), is not very satisfying since it involves the finding of certain substitutions that are not related to the m.g.u.'s. Fortunately, it follows from our results that there is an even simpler way to decide class M^+.

<u>DEFININITION</u> 5.28.: For any clause set S the set of depth-1-resolvents of S is defined as $R_{\leq 1}(S) = \{E \mid E \in R(S) \ \& \ \tau(E) \leq 1\}$.

In contrast to R_U, R_1 is an ordinary resolution refinement in the sense that $R_{\leq 1}(S) \subseteq R(S)$. Clearly, R_U-resolvents are instances of $R_{\leq 1}$ - resolvents. The following proposition expresses the fact that resolvents of instances are instances of resolvents (of the original clauses):

PROPOSITION 5.19.: Let $E \in R(\{C\sigma, D\vartheta\})$ then there exists an $E' \in R(\{C,D\})$ s.t. E is an instance of E'.

Thus we conclude that also $R_{\leq 1}$-resolution is complete on U if R_U is so. It then follows

THEOREM 5.20.: $R_{\leq 1}$ provides a decision procedure for M^+.

5.5.4 Completeness of R_U

In order to prove the (refutational) completeness of R_U for class U we make use of the concept of semantic trees. We assume the reader to be familiar with this important device (see e.g. [KH69], but briefly review the terminology.

For any clause set S the Herbrand base of S is the set of all ground atoms consisting of predicate, function symbols and constants occurring in S (If there are no constants in S we just introduce one). A semantic tree of S is a binary tree with elements of the Herbrand base and their negations labeling the edges. Let A_0, A_1, A_1, ... be an enumeration of the Herbrand base of S. The two edges leaving the root of the corresponding semantic tree T are labeled A_0 and $\neg A_0$, respectively; and if either A_i or $\neg A_i$ labels the edge entering a node, then A_{i+1} and $\neg A_{i+1}$ label the edges leaving that node. With each node of T we associate a refutation set, consisting of the literals labeling the edges of the path from the root down to this node.

A clause C in S fails at a node of T if the complement of every literal in some ground instance $C\gamma$ of C is in the refutation set of that node. A node k is a failure node for S if some clause of S fails at k but no clause of S fails on a node above k (i.e. a node with a shorter path to the root). A node is called inference node if both of its sons are failure nodes. T is closed for S if there is a failure node for S on every branch of T.

A completeness proof for a resolution strategy R' applied to clause sets S essentially rests on the following well known fact:

(1) If S is unsatisfiable, every semantic tree T for S is closed.

Our task is to show:

(2) There is a semantic tree T for S s.t. any two clauses which fail immediately below an inference node of T form an R'–resolvent which fails at the inference node.

Given (1) and (2) the completeness of R' follows by well known arguments (see e.g. [KH69]). Observe that T can be based on an arbitrary enumeration of the Herbrand base. Whereas any enumeration may be used to prove the completeness of Robinsons original resolution strategy, we will have to make a more judicious choice of the enumeration.

<u>THEOREM</u> S. 21.: Resolution procedure R_U is (refutationally) complete for all clause sets S in U.

<u>Proof.</u>
For any clause set S we define an enumeration A_0, A_1, ... of the corresponding Herbrand base s.t. deeper atoms succeed less deep ones. Within atoms of equal depth uniform atoms preceed those atoms that are not uniform. More formally we have:

(i) if $\tau(A_i) < \tau(A_j)$ then $i < j$, and
(ii) if $\tau(A_i) = \tau(A_j)$ and $i < j$ then A_j is uniform implies that A_i is uniform, too.

By the completeness of (Robinson's) resolution with respect to semantic trees it holds that, for any clauses C and D failing immediately below, but not at, a node k of T, there exists an R-resolvent E of C and D which fails at k. In the following ϑ denotes the m.g.u. of atoms in C and D used to resolve E. We show that, for C, D in U, there also exists an R_U-resolvent E' failing at node k.

We have to investigate the following cases:

(1) E is function free ($\tau(E) = 0$): Since any function free clause satisfies the defining conditions of U, E itself is the R_U-resolvent we are looking for.

(2) $\tau(E) > \max(\tau(C), \tau(D))$: We show that this cannot happen. By lemma S.17 the m.g.u. of two uniform, functional atoms of depth 1 is flat. Thus, the greater term depth of the resolvent can only arise if there is some function

free atom A among the atoms resolved upon and if $\vartheta(x)$ is functional for some $x \in V(A)$. Moreover there has to exist a functional atom B, not to be resolved upon but in the same clause as A s.t. x occurs in a functional term of B (i.e., $\tau_{max}(x,B) = 1$). Since $A\vartheta$ is uniform we have $V(A\vartheta) = V(\vartheta(x))$ and consequently $V(A\vartheta) \subseteq V(B\vartheta)$. Again by the uniformity of $A\vartheta$, it follows that

$$\tau(A\vartheta) < \tau(B\vartheta)$$

and thus also $\tau(A\vartheta\gamma) < \tau(E\gamma)$, where γ is the ground substitution s.t. $E\gamma$ is in the refutation set corresponding to node k. ($A\vartheta$ is the resolved atom). On the other hand the semantic tree is defined s.t. the ground instance of the resolved atom is at least as deep as all atoms in the refutation set at node k. This means that

$$\tau(A\vartheta\gamma) \geq \tau(E\gamma).$$

The contradiction implies that the resolution procedure remains complete if we discard resolvents that are deeper than their parent clauses.

By definition, $\max(\tau(C),\tau(D)) \leq 1$. It remains to investigate the following case:

(3) $\tau(E) = 1$: There are two subcases:

(3a) All literals in E are uniform: Then, like in case (1), E itself is an R_U-resolvent.

(3b) There are non-uniform literals in E: By lemma 5.17 ϑ is flat if in each parent clause there are functional literals among the literals resolved upon. If A is uniform then $A\vartheta$ is uniform, too, whenever ϑ is flat. Since not all literals of E are uniform it follows that all literals resolved upon in one of the parent clauses - w.l.o.g. say in C - are function free. Let M denote this subset of C. There has to exist some x in V(M) s.t. $\vartheta(x) = s$ is functional and x in V(B) for some function free B in C - M but $V(B\vartheta) \not\subseteq V(M\vartheta)$. (Since $V(s) = V(M\vartheta)$ we then know that $B\vartheta$ is a non - uniform literal of E.) Again, let γ be the ground substitution s.t. $E\gamma$ is in the refutation set corresponding to node k. By the definition of the enumeration of the Herbrand universe not only the single literal in $M\vartheta\gamma$, but all literals in $C\vartheta\gamma$ that are of depth $\tau(M\vartheta\gamma)$ are uniform; in particular also $B\vartheta\gamma$ is uniform. We shall specify a substitution σ s.t. $E\sigma\gamma = E\gamma$ and $B\vartheta\sigma$ is uniform and of depth ≤ 1.

Let $y \in V(B\vartheta) - V(s)$. The crucial fact is that (the only literal in) $M\vartheta\gamma$ and consequently also $B\vartheta\gamma$ is uniform on $s\gamma$. Therefore we know that either $\gamma(y) = c$ (for some constant c) or $\gamma(y) = t_i\gamma$, where t_i is an argument of s, or $\gamma(y) = s\gamma$. Corresponding to these cases we define $\sigma(y) = c$ or $\sigma(y) = t_i$ or $\sigma(y) = s$, respectively. Analogously we define $\sigma(z)$ for all variables z of $B\vartheta$ that do not occur in s. For all other variables let $\sigma(x) = x$. Clearly, $B\vartheta\sigma$ is uniform and of depth ≤ 1. We have to show that all other literals of $E\sigma$ remain of depth ≤ 1, too. Suppose $\tau(L\sigma) > 1$ for some L in E; then $\sigma(x) = s$ for some x that occurs as argument of a functional term in L. Since $\tau(s\gamma) = \tau(M\vartheta\gamma)$ we would then have $\tau(M\vartheta\gamma) < \tau(L\sigma\gamma)$. But by definition of the semantic tree we know that $\tau(M\vartheta\gamma) = \tau(E\gamma) = \tau(L\sigma\gamma)$. Therefore $\tau(E\sigma) = 1$.

There might still be literals in $E\sigma$ which are not yet uniform. For each such literal we proceed in the same way as for $B\vartheta$. Combining the corresponding substitutions we arrive at an instance $E\sigma'$ of E s.t. $E\sigma'\gamma = E\gamma$ but $\{E\sigma'\}$ in U.

Q.E.D.

It is interesting to observe that in the proof above we did not make use of the fact that the clause sets in M^+ are Krom. This implies that R_{M^+} remains complete for the class of clause sets defined by conditions (ii) and (iii) of definition 5.25 only. Of course this class is undecidable; nevertheless R_{M^+} provides a term depth limit for the resolvents. This emphasizes the significance of a length limit for the clauses.

Chapter 6

DECIDING THE CLASS K BY AN ORDERING REFINEMENT

6.1 THE CLASS K

Class K is a decidable class of formulas in first order locic described in [Mas 68]. It contains a lot of known decidable classe (e. g. all classes described in the chapter 2 of [DG 79] can be reduced to the class K).
Let us present the description of this class first in quantificational and then in its Skolemized form (as in [Zam 89]).

DEFINITION 6. 1.: The variables in the domain of a substitution λ are called proper variables of λ.

DEFINITION 6. 2.: Let F be a formula and G a subformula of F. The F - prefix of the formula G is a sequence of quantifiers of the formula F which bind the free variables of G.

The F - prefix is the F - prefix of any atomic subformula of the formula F.

DEFINITION 6. 3.: (The original definition of the class K was given by S. Maslov in a dual form). The formula F belongs to the class K in there exist variables $x_1, \ldots, x_k$, $k \geq 0$, which are not in the scope of any existential quanifier, such that each nonempty F - prefix:

- either is of length 1 (singular F - prefix),
- or ends with an existential quantifier,
- or is of the form $\forall x_1, \forall x_2, \ldots, \forall x_k$.

W. l. o. g. we can assume that every formula of the class K has the form

$$(\forall x_1)(\forall x_2) \ldots (\forall x_k)(Qy_1) \ldots (Qy_l) \mathop{\&}_{i=1}^{n} \bigvee_{j=1}^{m_i} A_{i,j} \qquad (1)$$

where $k \geq 0$, $l \geq 0$, $n > 0$, $m_i > 0$, $A_{i,j}$ are literals.

After Skolemisation the formula (I) will be in the for

$$\forall (\mathop{\&}_{i=1}^{n} \bigvee_{j=1}^{m_i} B_{i,j}) \qquad (2)$$

where each literal $B_{i,j}$ is obtained from $A_{i,j}$ by replacing $\exists$-variables and free variables by corresponding Skolem terms.
The class of formulas of the kind (2) obtained from formulas of the class K will be will denoted by KS.

DEFINITION 6.4.: A nonconstant argument of a literal is called essential and inessential otherwise.

Let us define some properties of formulas from the class KS referring to corresponding properties of formulas from the class K.

Table 1.

$F \in K$	$G \in KS$
F is a formula in prenex normal form	For any two functional G-terms t and s the argument list of t is the beginning of the arguments list of s or vice versa.
The quantifier $\exists x$ of a formula F is in the scope of a quantifier $\forall y$	The Skolem G-term replacing the variable x contains the variable y as an argument
F-prefix	The set of essential arguments of the corresponding atom contains a functional term t such that every variable of this set is an argument of t, and the list of arguments of any other functional term from this set is a beginning of the list of arguments of the term t

The F-prefix has the form $\forall x_1 \dots \forall x_k$	The set of essential arguments consists of variables $x_1, \dots, x_k$
The F-prefix has a length 1	The set of essential arguments is a singleton set.

DEFINITION 6.5.: The term t dominates the term s if at least one of the following conditions holds:

(1) $t = s$,

(2) t is a functional term and s is a variable argument of t,

(3) $t = f(t_1, \dots, t_n)$, $s = g(t_1, \dots, t_m)$, $n \geq m \geq 0$.

DEFINITION 6.6.: The set T_1 of terms dominates the set T_2 of terms if for every term t_2 from T_2 there exists a term t_1 from T_1 such that t_1 dominates t_2.

DEFINITION 6.7.: The literal L_1 dominates the literal L_2 if the set of arguments of L_1 dominates the set of arguments of L_2

DEFINITION 6.8.: Terms t and s are similar if t dominates s and s dominates t. The similarity relation for literals and term sets is defined analogously.

EXAMPLE 6.1.: The term $f(x, y, g(x))$ dominates the terms $x, y, g(x)$, $h(x,y)$, $h(x, y, g(x))$. The literal $A(x, y, f(x,y))$ dominates the literals $B(y)$, $C(g(x), y)$, $D(y, h(x,y), x)$. The terms $f(x, y, g(x, y))$ and $h(x, y, g(x, y))$ are similar but $f(x, y, g(x, y))$ and $h(y, x, g(x, y))$ are not. The literals $A(x, y, g(x))$, $A(y, x, g(x))$, $B(x,y,x,y, h(x))$ are similar.

DEFINITION 6.9.: A term is called regular if it dominates all its arguments.

In particular, variables and constants are trivially regular, all terms of depth 1 are regular.

EXAMPLE 6.2.: The terms x, f(x,y,z), f(x,g(x),z), f(x,g(x), h(x,g(x))) are regular. The term f(x,h(x,g(x))) is not regular since it does not dominate its second argument.

We can see from the definition of a regular term, that it is possible to use a short form of any regular term:
we may omit the arguments of leading functional subterms, since the position of any function symbol completely determines its arguments (it is assumed that the arities of all functional symbols in a short from are known).

EXAMPLE 6.3.: The short forms of the last two regular terms from the example above are f(x,g,z), f(x,g,h).

DEFINITION 6.10.: A set of terms is called regular if it contains no functional terms or it contains some regular functional term which dominates all terms of this set. (We will call such a term a dominating term for this set).

DEFINITION 6.11.: A literal is called regular if the set of its arguments is regular.

EXAMPLE 6.4.: The literals A(x,y,f(x,y), g(x)), B(x,f(x,g(x))), C(x,a,y) are regular.

DEFINITION 6.12.: A literal of depth 0 is called unary if the set of its essential arguments consists of only one variable.

Some obvious properties of regular terms and of the dominating and similarity relations are given in the form of the following remarks.

Remark 1. If the term t dominates the terms t_1 and t_2 and the arity of the term t_1 is not less than the arity of the term t_2 then t_1 dominates t_2.

Remark 2. The dominating relation is transitive and, for functional terms, it is preserved under substitution.

Remark 3. The similarity relation is transitive.

Remark 4. Sets of essential arguments of any two similar literals of depth 0 are equal.

DEFINITION 6.13.: A set M of literals is called k-regular if the following conditions hold:

1) M contains regular literals only,
2) the nonnegative integer k is not greater than the minimum arity of function symbols occurring in literals of the set M,
3) M contains some literal which dominates every literal from the set M,
4) All nonunary literals of depth 0 are similar and the set of essential arguments of any such literal is similar to the set of the first k arguments of any functional term occuring in any literal from M.

A set M of literals is called regular if it is k-regular for some k, $k \geq 0$.

DEFINITION 6.14.: A regular literal of positive depth called a literal of type 1, a nonunary literal of depth zero is called a literal of type 2, an unary literal of depth zero is called a literal of type 3.

EXAMPLE 6.5.: The set of literals $\{A(x, z, f(x, y, z)), B(x, g(x, y)), C(x, y), D(z)\}$ is 2-regular.

DEFINITION 6.15.: A subset M_1 of the set M of literals is called a component of M if M_1 and $H - M_1$ are variable disjoint. A set M is called decomposable if it has at least two nonempty components and indecomposable otherwise.

DEFINITION 6.16.: A set of literals is called quasiregular if all of its indecomposable components are regular.

<u>THEOREM</u> 6.1.: Every clause C of a formula of the form (2) is quasiregular.

<u>Proof</u>:

We first show that C consists of regular literals only. Let A be a literal from C and let B be the corresponding literal in a formula of type (1). We will consider the following cases:

1) The F-prefix of the literal B ends with $\exists y$. Then the literal A contains the term t (replacing the variable y after Skolemization) which dominates all arguments of the literal A.
2) The F-prefix of B is of the form $\forall x_1 \ldots \forall x_k$. In this case B does not contain functional terms and therefore is regular.
3) The F-prefix of B consists of one $\forall$-quantifier. In this case B is regular too, since it does not contain any functional terms.
4) The F-prefix of B is empty. Similar to cases 1) and 2).

Now let us show that C is quasiregular. Let C_1 be an indecomposable component of C. Since C_1 is indecomposable it either consists of one literal - in this case C_1 is trivially regular - or it does not contain any literals. Let us consider the following cases:

1) Let A be a literal of positive depth from C_1 such that its dominating term t_1 has a maximum arity of all terms occurring in C_1. Let B be another literal from C_1. If B contains a functional term t_2 then the arity of t_1 is not less then the arity of t_2. Since F is a prenex formula and t_1, t_2 are Skolem terms, the argument list of t_2 is a beginning of the argument list of the term t_1, i.e. t_1 dominates t_2. It follows that A dominates B. If the literal B does not contain functional terms and B is not unary then the set of its essential arguments is equal to the set of the first k arguments of the term t_1, therefore A dominates B. Finally if B is a unary literal and does not contain function symbols then its only variable argument is an argument of t_1, since C_1 is indecomposable. Therefore A dominates B.

2) Suppose that C_1 does not contain literals with functional terms. Then every literal occurring in C_1 is obtained from some literal with F-prefix $\forall x_1 \ldots \forall x_k$ or $\forall z$ for some z. Let A and B be literals from C_1 where A is obtained from

a literal with F-prefix $\forall x_1 \ldots \forall x_k$. If B is obtained from a literal with the same F-prefix, then A and B are similar. If B is obtained from a literal with F-prefix $\forall z$ then the only variable of B is x_i for some i, $1 \leq i \leq k$, since C_1 is indecomposable. So in both cases A dominates B.

3) If C_1 consists of literals which are obtained from literals with singular F-prefix then, since C_1 is indecomposable, all literals contain the same variable. If follows that all literals in C_1 are similar.

So we habe shown that in any case C_1 contains some literal which dominates all literals from C_1, i.e. C_1 is a regular component.

Q.E.D.

REGULAR TERMS AND CLAUSES

In the following lemmas we investigate properties of regular terms and clauses.

LEMMA 6.1.: A term t is regular iff it dominates each of its arguments of positive depth.

Proof:

The "if" part follows from the fact that every functional term dominates every constant and every variable argument. The "only if" part is evident.

LEMMA 6.2.: If t is a regular term and t dominates a term s then s is regular too.

Proof:

If $\tau(s) = 0$, then s is trivially regular. Let s be a functional term of the form $g(s_1, \ldots, s_m)$. Since t dominates s, it follows that t is a functional term of the kind $f(s_1, \ldots, s_m, \ldots, s_n)$, where f and g are functional symbols, $s_1, \ldots, s_n$ are terms, $n \geq m > 0$. By lemma 6.1 it is enough to prove that s dominates each of its arguments of positive depth. Let $s_i = h(p_1, \ldots, p_e)$ be some argument of the term s, $i \leq m$. Since t is regular, it dominates s_i and it follows that $p_1 = s_1, \ldots, p_e = s_e$. Moreover we have $e < i$, since otherwise $p_i = s_i$, i.e. s_i is equal to one of its argument, which is impossible. Since $e < i$ and $i \leq m$ it holds $e < m$ and each argument of the term s_i is equal to a corresponding argument of the term s. Therefore s dominates s_i which had to be proved.

Corollary 6.1.: Every subterm of a regular term is regular.

Corollary 6.2.: Every regular term dominates each of its subterms.

In general the regularity of terms is not preserved by substitutions. For example, if we substitute the regular term g(x) into a regular term f(y, z) for the variable y then we will obtain the irregular term f(g(x), z). However, a substitution which unifies some set of regular terms preserves regularity. The proof of this fact is given below.

LEMMA 6.3.: Let t be a regular term, and $\sigma = \{t_1/x_1, \dots, t_n/x_n\}$ be a substitution, where $t_1, \dots, t_n$ are constants or variables; then $(t)\sigma$ is a regular term.

Proof:
The substitution σ preserves the dominating relation for terms occurring in t as well as the depth of terms.

LEMMA 6.4.: Let the regular term $t = f(t_1, \dots, t_n)$ dominate the term $s = g(t_1, \dots, t_m)$, $n \geq m \geq 1$, and let σ be a substitution such that all its proper variables occur in s or do not occur in t. If $(s)\sigma$ is regular, then $(t)\sigma$ is regular too.

Proof:
It is sufficient to prove that $(t)\sigma$ dominates each of its argument of positive depth (by lemma 6.1). Let $(t_j)\sigma$ be a term of positive depth, $1 \leq j \leq n$. We will consider two cases:

1) t_j is a variable. Then t_j is a proper variable of the substitution σ, since otherwise $(t_j)\sigma = t_j$ which is impossible because $(t_j)\sigma$ is of positive depth. By the assumption of the lemma, t_j occurs in s. The term $(t_j)\sigma$ is a subterm of the term $(s)\sigma$ therefore $(s)\sigma$ dominates $(t_j)\sigma$ (since $(s)\sigma$ is regular). By remark 2 the dominating relation is transitive and is preserved under substituting terms of positive depth. Since t dominates s, $(t)\sigma$ dominates $(s)\sigma$, therefore $(t)\sigma$ dominates $(t_j)\sigma$.

2) t_j is a term of positive depth. It follows that t dominates t_j, since t is regular. Since the dominating relation is preserved under substitution in terms of positive depth, the term $(t)\sigma$ dominates $(t_j)\sigma$.

So we have shown that $(t)\sigma$ dominates each its argument of positive depth and thus is regular.

LEMMA 6.5.: Let $t = f(p_1, \dots, p_n)$ be a regular term which dominates the term $s = g(p_1, \dots, p_m)$, z be a variable which does not occur in s and let σ be the substitution $\{s/z\}$. Then $(t)\sigma$ is regular.

Proof:

Let u be an argument of t and let the depth of $(u)\sigma$ be positive. There are the following cases:

1) u is a variable different from z. Then $(u)\sigma = 0$ which is impossible since the depth of u is equal to 0 but the depth of $(u)\sigma$ is positive.

2) u is equal to z. In this case $(u)\sigma = s$. Since s does not contain z, the terms $p_1, \dots, p_m$ do not contain z too. Therefore

$$(t)\sigma = f(p_1, \dots, p_m, p_{m+1}, \dots, p_n)\sigma = f((p_1)\sigma, \dots, (p_m)\sigma, \dots, (p_n)\sigma) =$$
$$= f(p_1, \dots, p_m, (p_{m+1})\sigma, \dots, (p_n)\sigma).$$

It follows that $(t)\sigma$ dominates $(u)\sigma$.

3) u is a term of positive depth or a constant. Since t is regular, it dominates the term u, therefore $(t)\sigma$ dominates $(u)\sigma$ (according to remark 2).

LEMMA 6.6.: If t and s are regular terms and σ is a m.g.u. for $\{t, s\}$ then $(t)\sigma$ is regular.

Proof:

Let us consider the following cases:

1) One of the terms (s, for example) is a variable. Then t does not contain s, $\sigma = \{t/s\}$ and $(t)\sigma = t$ which is regular.

2) Both of t, s are functional terms of the kind $f(p_1, \dots, p_n)$ and $f(q_1, \dots, q_n)$ respectively (they begin with the same function symbol f because they are unifiable). We will prove the regularity of $(t)\sigma$ by induction on $(n-i)$, where i satisfies the following conditions:

$$p_1 = q_1, \dots, p_i = q_i, \; p_{i+1} \neq q_{i+1}, \; (1 \le i < n) \qquad (1)$$

Since t and s are unifiable, then terms p_{i+1} and q_{i+1} are unifiable too.
The induction base is trivial.
Let $n-i > 0$. In this case one of the terms p_{i+1}, q_{i+1} must be a variable. Assume the contrary and consider the following subcases:

2.1) The depth of both terms is zero. In this case terms p_{i+1} and q_{i+1} are constants. They are different by the condition (1); but then they are not unifiable which contradicts with unifiablility of t and s.

2.2) The depth of p_{i+1} is positive and the depth of s is zero. In this case these terms are not unifiable too.

2.3) Both terms are of positive depth. Assume that $p_{i+1} = g(u_1, \dots, u_m)$, $q_{i+1} = g(v_1, \dots, v_m)$. Since t is regular, it dominates the term p_{i+1}, therefore

$$p_1 = u_1, \dots, p_m = u_m \qquad (2)$$

Since no term contains itself as argument, it follows that $p_{i+1} \neq u_k$ for each k, $k = 1, \dots, m$. Therefore $i + 1 > m$.
Similarly we can show that

$$q_1 = v_1, \dots, q_m = v_m. \qquad (3)$$

So we can conclude from (1) - (3) that $u_k = p_k = q_k = v_k$ for every k, $k = 1, \dots m$. It follows that $p_{i+1} = q_{i+1}$ which contradicts (1).

Thus we have proved that at least one of the terms p_{i+1} or q_{i+1} (say the first one) is a variable. Let σ_{i+1} be the substitution $\{q_{i+1}/p_{i+1}\}$. If q_{i+1} is a variable, then both $(t)\sigma$ and $(s)\sigma$ are regular by lemma 3.

Let q_{i+1} be of positive depth (it does not contain p_{i+1}). Since s is regular, it dominates the term q_{i+1}. Since the first i arguments of terms t and s are equal by (1), it follows that t dominates q_{i+1} too. By lemma 6.5 the terms

$(t)\sigma_{i+1}$ and $(s)\sigma_{i+1}$ are regular. It follows that the following conditions for the arguments of $(t)\sigma_{i+1}$ and $(s)\sigma_{i+1}$ are valid:

$$(p_1) \quad \sigma_{i+1} = (q_1)\sigma_{i+1}, \ldots$$
$$(p_1) \quad \sigma_{i+1} = (q_i)\sigma_{i+1},$$
$$(p_{i+1}) \quad \sigma_{i+1} = (q_{i+1})\sigma_{i+1} \ldots$$

By the induction hypothesis $((t)\sigma_{i+1})\lambda$ and $((s)\sigma_{i+1})\lambda$ are regular terms, where λ is a m.g.u. for $((t)\sigma_{i+1})$ and $((s)\sigma_{i+1}$.

Q.E.D.

Let us denote by ar(t) the arity of the term t and by minar(t) the minimum arity of nonconstant function symbols occurring in the term t. If t does not contain nonconstant function symbols then let minar(t) = 0.

THEOREM 6.2.: The depth of any regular term t is less or equal to ar(t) - minar(t) + 1.

Proof:
By induction on $\tau(t)$:
For $\tau(t) = 0$ the theorem obviously holds. Let us denote by s the argument of t with maximum depth. It is obvious that

$$\tau(s) + 1 = \tau(t) \text{ and } ar(s) < ar(t) \qquad (4)$$

By corollary 6.1 s is a regular term and by the induction hypothesis it holds $\tau(s) \leq ar(s) - minar(s) + 1$. It follows by (4) that

$$\tau(t) = \tau(s) + 1 \leq ar(s) + 1 - minar(s) + 1 \leq ar(t) - minar(t) + 1,$$

which had to be proved.

Q.E.D.

Now we will prove some properties of regular clauses.

LEMMA 6.7.: Let $A_1 = A(t_1, \ldots, t_n)$ and $A_2 = A(p_1, \ldots, p_n)$ be unifiable literals of type 1. If t_i is a dominating term of A_1, then also p_i must be a dominating term of A_2.

Proof:

Since A_1 is a literal of type 1, its dominating term has maximal arity among other arguments of this literal and is of positive depth. Let p_j be a dominating term of a literal A_2, i. e. p_j is a term of positive depth whose arity is maximal among arities of all arguments of A_2. Let us denote by σ the m.g.u. of A_1 and A_2. There are the following cases:

1) Terms p_i and t_j are variables. Since A_1 is a literal of type 1, the variable t_j is an argument of the dominating term t_i. For the same reason p_i is an argument of the term p_j. The relation "to be an argument" is preserved under substitution, therefore the following statements hold:

   $(t_j)\sigma$ is an argument of $(t_i)\sigma$ and
   $(p_i)\sigma$ is an argument of $(p_j)\sigma$ (5)

   From the unifiability of A_1 and A_2 we conclude that

   $(t_i)\sigma = (p_i)\sigma$ and $(t_j)\sigma = (p_j)\sigma$ (6)

   It follows from (5) and (6) that $(t_j)\sigma$ is an argument of an own argument (and thus a proper subterm of itself), which is impossible.

2) One of the terms t_j or p_i (say, t_j) is a variable and the other one is a constant or is of positive depth. Then t_j is an argument of t_i and p_j dominates the term p_i, therefore $ar(p_i) \le ar(p_j)$. Since, for non-variable terms, the dominating relation is preserved under substitution, the following statements hold:

   $(t_j)\sigma$ is an argument of $(t_i)\sigma$ and
   $(p_j)\sigma$ dominates $(p_i)\sigma$ (7)

   From (6) and (7) we conclude that $(t_j)\sigma$ is an argument of some term which is dominated by $(p_j)\sigma$. Therefore $(t_j)\sigma$ is an argument of $(p_j)\sigma$, which is impossible.

3) Both terms p_i and t_j are not variables. Then for their arities we get:

   $ar(t_i) \ge ar(t_j)$ since t_i is a dominating term for A_1
   $ar(p_j) \ge ar(p_i)$ since p_j is a dominating term for A_2
   $ar(t_i) = ar(p_i)$ since t_i and p_i are unifiable,
   $ar(t_j) = ar(p_j)$ since t_j and p_j are unifiable.

These statements imply that $ar(t_i) = ar(t_j) = ar(p_i) = ar(p_j)$. It follows that p_i is a dominating term for A_2 too.

Q.E.D.

LEMMA 6.8.: Assume that A is a regular literal of the type 1, the term t is a dominating term of A and α is a substitution such that $(t)\alpha$ is regular. Then $(A)\alpha$ is a regular literal.

Proof:
Let us show that $(t)\alpha$ dominates each argument of $(A)\alpha$. Assume that u is some argument of A. There are the following cases:

1) u is a variable. In this case u is an argument of the dominating term t since A is a regular literal. It follows that $(u)\alpha$ is an argument of $(t)\alpha$. Since $(t)\alpha$ is a regular term, it dominates each of its arguments, particularly the term $(u)\alpha$.

2) u is a constant or a term of positive depth. Then t dominates u. In this case the dominating relation is preserved under substitution. Therefore $(t)\alpha$ dominates the term $(u)\alpha$.

Q.E.D.

LEMMA 6.9.: Assume that $A_1 = A(t_1, \dots, t_n)$ and $A_2 = A(p_1, \dots, p_n)$ are regular literals and α is a m.g.u. of A_1 and A_2. Then $(A_1)\alpha$ is regular.

Proof:
We will consider the following cases:

1) A_1 and A_2 are both literals of type 1. Assume that t_i is a dominating term for A_1. By lemma 6.6 the dominating term for A_2 is the term p_i. Let Θ be the m.g.u. for t_i and p_i. By lemma 6.6 the terms $(t_i)\Theta$ and $(p_i)\Theta$ are regular. By lemma 6.8 the literals $(A_1)\Theta$ $(A_2)\Theta$ are regular and their dominating terms $(t_i)\Theta$ and $(p_i)\Theta$ are equal.

   Now we use induction on the number of nonequal arguments in $(A_1)\Theta$ and $(A_2)\Theta$ to prove the lemma.
   The induction base is trivial. Let $(t_j)\Theta$ differ from $(p_j)\Theta$ for some j, $1 \le j \le n$,

and let α_j be a m.g.u. for these terms. Since $(t_i)\Theta$ is regular and dominates the term $(t_j)\Theta$, this last term is regular too by lemma 6.2. For the same reason also $(p_j)\Theta$ is regular. The term $((t_i)\Theta)\alpha_j$ is regular by lemma 6.4. Therefore the literal $((A_1)\Theta)\alpha_j$ is regular by lemma 6.8. Similarly we can prove that the literal $((A_2)\Theta)\alpha_j$ is regular too. By the induction hypothesis the lemma holds for the literals $((A_1)\Theta)\alpha_j$ and $((A_2)\Theta)\alpha_j$.

2) One of the literals, A_2 for example, is of term depth 0 and the other one is a literal of the type 1 with a dominating term t_i. In this case we can prove the lemma by induction on the number of pairs of nonequal terms (t_e, t_j) where e and j are numbers with $p_e = p_j$, $1 \leq e, j \leq n$.
The induction base is trivial because the unifier is a match.

Assume that $p_e = p_j$ and $t_e \neq t_j$. Since A_1 and A_2 are unifiable t_e and t_j are unifiable too. Let μ be a m.g.u. for t_e and t_j. By lemma 6.4 the term $(t_i)\mu$ is regular. By lemma 6.8. the literal $(A_1)\mu$ is regular. For the literals $(A_1)\mu$ and $(A_2)\mu$ the lemma holds by the induction hypothesis.

3) The literals A_1 and A_2 are both of term depth zero. The proof is evident since $(A_1)\sigma$ and $(A_2)\sigma$ do not contain functional terms and therefore are regular.

Q.E.D.

<u>LEMMA</u> 6.10.: Let M be a regular set of literals, A be a dominating literal for M and t be a dominating term for A. Then t dominates each argument of each literal from M.

<u>Proof</u>:
Let s be some argument of some literal B from M. Since A is a dominating literal, it dominates B i.e. there exists some term in A which dominates s. Since A is regular, t dominates each argument of A, therefore dominates s, which was to show.

Q.E.D.

<u>LEMMA</u> 6.11.: If a regular term t is a dominating term for a regular set M of literals, then every literal A which contains t is a dominating literal for M.

The proof follows immediately from the definition of the dominating relation for literals.

LEMMA 6.12.: If a regular term t dominates the term s and σ is a substitution such that $(t)\sigma$ is regular, then $(t)\sigma$ dominates $(s)\sigma$.

Proof:
The case t = s is trivial. Suppose now that t and s are different. Since t dominates s, the depth of t is nonzero. Assume that the depth of s is zero. If s is constant, the proof is evident. Let s be a variable. Then it is an argument of the term t. The relation "to be an argument" is preserved substitution, therefore $(s)\sigma$ is an argument of $(t)\sigma$. Since $(t)\sigma$ is regular, it dominates each of its arguments, in particular it dominates $(s)\sigma$. Assume now that the depth of s is greater than zero. Since the dominating relation is preserved under substitution in terms of positive depth, $(t)\sigma$ dominates $(s)\sigma$.

Q.E.D.

LEMMA 6.13.: Let F be a formula, t be a regular F-term and let k be the minimal arity of function symbols occurring in F. Then the first k arguments of the term t are of depth 0.

Proof.
Let $t = f(t_1, \dots, t_n)$, $n \geq k$. Assume that the depth of t_i is nonzero for some i, $1 \leq i \leq k$, i.e. $t_i = g(u_1, \dots, u_m)$ for some m, $k \leq m \leq n$. Since the term t is regular, it dominates t_i, therefore $u_j = t_j$ for every j, $1 \leq j \leq m$. It follows that $t_i = g(t_1, \dots, t_m)$ for some i, $1 \leq i \leq k \leq m$, which is impossible.

Q.E.D.

LEMMA 6.14.: Assume that {A, B} is a k-regular indecomposable set of F-literals for a formula F, such that A dominates B and σ is a substitution such that $(A)\sigma$ is regular; suppose further that k is not greater than the minimal arity of function symbols occurring in $(A)\sigma$. Then $(A)\sigma$ dominates $(B)\sigma$ and the set $\{A, B\}\sigma$ is k-regular.

Proof.
Let us consider following cases:

1) Both A and B are literals of type 1.

Since A dominates B, the dominating term t_1 of the literal A dominates each argument of the literal B by lemma 6.10. Since the literal $(A)\sigma$ is regular, the term $(t_1)\sigma$ is regular and dominates each argument of the literal $(B)\sigma$ by lemma 6.12. It follows that $(A)\sigma$ dominates $(B)\sigma$. Now let us prove that $(B)\sigma$ is a regular literal. The term t_1 dominates the dominating term t_2 of the literal B by lemma 6.10., therefore $(t_1)\sigma$ dominates $(t_2)\sigma$ by lemma 6.12. Since $(t_1)\sigma$ is regular, $(t_2)\sigma$ is regular too by lemma 6.2; moreover the term $(t_2)\sigma$ dominates each argument of the literal $(B)\sigma$ by lemma 6.12. It follows that $(B)\sigma$ is regular. So we have shown that $(A)\sigma$ dominates $(B)\sigma$ and both literals are regular, i.e. the set $\{(A)\sigma, (B)\sigma\}$ is a k-regular set by definition.

2) A is a literal of type 1, B is a literal of type 2.

Then the set of arguments of the literal B is similar to the set consisting of the first k arguments of the term t_1. Since $(A)\sigma$ is regular, $(t_1)\sigma$ is regular and its first k arguments are terms of depth zero by lemma 6.13. Therefore the set of arguments of $(B)\sigma$ consists of the first k arguments of the term $(t_1)\sigma$ and may be of constants. It follows that $(B)\sigma$ is regular and $\{(A)\sigma, (B)\sigma\}$ is a k-regular set.

3) A is a literal of type 1. B is aliteral of the type 3.

Let x be the only variable occurring in B. Then x is an argument of the dominating term t_1 of the literal A since $\{A, B\}$ is an indecomposable set. Since $(t_1)\sigma$ is regular, it dominates $(x)\sigma$ by lemma 6.12. Therefore $(x)\sigma$ is regular, which implies that $(B)\sigma$ is regular. So we have shown that $(A)\sigma$ dominates $(B)\sigma$ and both literals are regular, i.e. the set $\{(A)\sigma, (B)\sigma\}$ is a k-regular set by defintion of k-regularity.

4) Both A and B are literals of type 2.

Since $\{A, B\}$ is regular, A and B are similar and the sets of their arguments differ in constants only. Since $(A)\sigma$ is regular. $(B)\sigma$ is regular too. So we have shown that $\{A, B\}\sigma$ is a k-regular set.

5) A is a literal of type 2, B is a literal of type 3.

In this case the variable x, occurring in B, occurs in A too, since $\{A,B\}$ is indecomposable. The literal $(A)\sigma$ is regular, therfore $(x)\sigma$ is regular too. It follows that the literal $(B)\sigma$ is regular. It is also obvious that $(A)\sigma$ dominates $(B)\sigma$.

6) Both A and B are literals of type 3. Similarly to the previous case.

Q.E.D.

Corollary 6. 3.: If M is a regular indecomposable set of literals, A is a dominating literal and $(A)\sigma$ is regular, then $(M)\sigma$ is regular.

LEMMA 6. 15.: Assume that $\{A, B\}$ is a regular set of literals, A is a literal of type 1 and σ is a substitution such that $A\sigma$ regular and each of its proper variables either occurs in A or does not occur in B. Then $\{A, B\}\sigma$ is regular.

Proof:
The case where A dominates B is considered in lemma 6. 14. Let B dominate A. It means that the dominating term t_2 of the literal B dominates the dominating term t_1 of the literal A. Since $(A)\sigma$ is regular, $(t_1)\sigma$ is regular too. By lemma 6. 4 the term $(t_2)\sigma$ is regular. By lemma 6. 8 $(B)\sigma$ is a regular literal. Since the dominating relation for terms of positive depth terms is preserved under substitution, $(t_2)\sigma$ dominates $(t_1)\sigma$. It follows that $(B)\sigma$ dominates $(A)\sigma$, i.e. $\{A, B\}\sigma$ is a regular set, which was to show.

Q.E.D.

LEMMA 6. 16.: Assume that $(C_1 \cup \{B\})$ and $(C_2 \cup \{\neg B\})$ are indecomposable, regular clauses and that the clause $C_1 \cup C_2$ is the resolvent of these clauses with empty m.g.u.. Then if the literals B and $\neg B$ are dominating literals for their clauses, then $C_1 \cup C_2$ is quasiregular.

Proof:
From the assumptions of the lemma it follows that B dominates C_2 as well. Therefore the set M of all literals which occur in the clause $\{B\} \cup C_1 \cup C_2$ is regular. Now we will prove that the set M_1 obtained from M by removing B is quasiregular.
Let M_1 contain some literal of type 1 and A be a literal containing a term with maximum arity. The literal A is regular since it occurs in a regular set. Let M_2

be the set of all literals from M_1 which contain no variable from A. Then A is a dominating literal for $M_1 - M_2$. In fact, the dominating term t_1 for A dominates each term of any literal of types 1 and 2 since it has a maximum arity among all terms of the set $M_1 - M_2$; by lemma 6.1 t_1 also dominates each argument of any literal of type 3 from $M_1 - M_2$. The set M_2 consists of literals of type 3 only and therefore is quasiregular.

In the case where M_1 contains no literal of type 1, but contains some literal of type 2, we can take A to be a literal of type 2. If M_1 consists of literals of the type 3 only, we can take an arbitrary literal from M_1 for A. So we have shown that M_1 is decomposable in two components M_2 and $M_1 - M_2$, one of which is regular, and the second one is quasiregular, which was to show.

Q.E.D.

LEMMA 6.17.: Assume that $C_1 \cup \{A\}$ and $C_2 \cup \{B\}$ are regular indecomposable clauses and the clause $(C_1 \cup C_2)\sigma$ is the resolvent of these clauses by resolution upon A and B. Then if the literals α and B are dominating for their clauses, then $(C_1 \cup C_2)\sigma$ is quasiregular.

Proof:

According to lemma 6.9 the literals $A\sigma$, $B\sigma$ are regular. By lemma 6.14 the clauses $(C_1 \cup \{A\})\sigma$ and $(C_2 \cup \{B\})\sigma$ are quasiregular (note that $A^d\sigma = B\sigma$). We conlude that the clause $(C_1 \cup C_2)\sigma$ is quasiregular by lemma 6.16.

Q.E.D.

Corollary 6.4.: The lemma holds if holds if the parents of the resolvent are quasiregular.

6.2 A DECISION PROCEDURE FOR THE CLASS KS

We will use a π - ordering refinement (see chapter 4) to decide KS. Recall that for π - orderings we consider clauses as lists rather than sets of literals. Here we define a specific π - ordering in the following way:

Let F be a formula from KS and HF its Herbrand expansion. Let the term list $t_1, \dots, t_n, \dots$ be a list of all terms from HF satisfying the following conditions:

1) If $\tau(t_j) < \tau(t_i)$ then $i > j$.
2) If t_i dominates t_j then $i > j$.
3) If the terms t_i and t_j are similar (i.e. they dominate each other) then the ordering is lexicographical.

Now we can define the ordering $L_1, L_2, \dots, L_n, \dots$ for all occurrences of literals in HF in the following way:

1) If the maximal term (in the sense of the term ordering above) of L_i is greater than the maximal term of L_j then $i > j$.
2) If the maximal terms of L_i and L_j are equal and the set of essential arguments of L_j is a proper subset of the analogous set of L_i then $i > j$.
3) If the sets of essential arguments of literals L_i and L_j are equal, L_i is an instance of some literal of type 2 and L_j is an instance of some literal of type 3, then $i > j$.

Any clause $L_{i_1}, \dots, L_{i_k}$ of literals from the ordered list $L_1, \dots, L_n, \dots$ will be called π - ordered if for all $e, m \in N$ $e < m$ implies $L_e < L_m$. If the formula F is unsatisfiable then (by Herbrand's theorem) there exists a finite part HF' of HF which is unsatisfiable as well.

According to the completeness of π - ordering refinements (proved in chapter 4) there exists a π - ordered refutation R of HF' (i.e. there exists a list of π - ordered clauses, each of which is either in HF' or derived from HF' by the π-ordering refinement of resolution). Note that, in this case, the acyclicity of the π-ordering is quaranteed by the fact that HF' is propositional.

Let R' be the lifted refutation obtained from R and let σ be the substituion with $R'\sigma = R$. We then have:

<u>LEMMA</u> 6.18.: Let C be a clause in the refutation R'. If C is regular then the resolved literal is dominating for C.

Proof:

We assume to the contrary that the resolved literal A from C is not dominating and that B is a dominating literal of C. It follows that there exists some term t_B in B which is not dominated by any term in A. Let us consider the following cases:

1) The literal B contains a functional term.
Then its dominating term t_B dominates each term in A but is not dominated by any term in A. It follows that each term in $A\sigma$ either is dominated by $t_B\sigma$ or is a proper subterm of $t_B\sigma$. In this case the literal $B\sigma$ must be greater than $A\sigma$ according to point 1) of the definition of literal ordering, but this contradicts the fact that A is the resolved literal.

2) The literal B does not contain function symbols.
Then there exists a variable x s.t. $x \in V(B) - V(A)$. If follows that B is of type 2 and A is of type 3. If $B\sigma$ contains at least one essential term which does not occur in $A\sigma$ then $B\sigma$ must be greater than $A\sigma$, according to point 2) of the literal ordering and we get a contradiction. If the sets of essential arguments of$A\sigma$ and $B\sigma$ are equal then $B\sigma$ must be greater than $A\sigma$, according to point 3) of the definition of the literal ordering; again we get a contradiction.

Thus we have shown that the resolved literal in a regular clause must be dominating.

Q.E.D.

THEOREM 6.3.: The class KS is decidable.

Proof:

Let F be a formula in KS. We have shown that, in case F is unsatisfiable, there exists a π-refutation of HF. Using induction on the length of R we can prove that every clause in a lifted refutation R' is quasiregular (by theorem 6.1 and lemma 6.17). It remains to show that the set of clauses which are π-derivable from a set of clauses in KS is always finite. Let S be a set of clauses in KS and $C \in S$. It is obvious that the number of variables in C does not exceed the number of universal quantified variables in the prefix of the K-formula corresponding to S.

Let ϑ be the number of variables, m the maximal arity of function symbols and φ be the number of different function symbols of the formula F. Then the number r of all regular F-terms cannot exceed $\varphi(\vartheta + \varphi)^m$. Indeed, if we consider the short form of a regular term we can choose one of φ function symbols, and one of ϑ variables or one of φ function symbols for each of m arguments.

Now let p be the number of different predicate symbols and n be the maximal arity of predicate symbols in the formula F. Then the number of atoms in a regular clause cannot exceed $p \cdot r^n$. Indeed we can choose one of p predicate symbols as leading predicate symbol and any of r regular terms for each of the n arguments. Thus the maximal length of a regular clause (corresponding to F) is less or equal to $p \cdot r^n$.

Let $\alpha = p \cdot r^n$; then the number c of regular clauses cannot exceed 3^α. Thus c is the maximal number of literals in an indecomposable component of a quasiregular clause and consequently the number of quasiregular clauses is less or equal 3^c. We conclude that the number of π-derivable clauses is finite.

Q.E.D.

It is easy to show that the class defined by building conjunctions of formulas from K is decidable (in the same way).

It was shown in lemma 6.18 that lifting a ground refutation (based on the specific ground ordering defined in this section) the resolved literals on the general level are always dominating. Thus we always find a refutation where no strictly dominated literal is resolved. Formally we may define the following ordering:

DEFINITION 6.17.: A $>_k$ B iff A dominates B, but B does not dominate A.

It is interesting to note that $>_k$ itself is not a π-ordering because property (D) from chapter 4 is not fulfilled (for a detailed discussion of such orderings see chapter 4.4).

We thus get the following resolution decision procedure for K:

a) Split the set of clauses into sets of sets of regular clauses.

b) Apply the $>_k$-ordering refinement to each set of clauses.

By completeness and termination of the $>_k$- refinement (on sets of regular clauses) a), b) indeed gives a decision procedure for K.

Chapter 7

A RESOLUTION BASED METHOD FOR BUILDING FINITE MODELS

In the following we will present an algorithm for building finite clause sets for formulas in the class we call AM, which is the union of the Initially-extended Ackermann and Essentially Monadic Classes. The initially extended Ackermann's Class is the class of formulas with a prefix $\exists^* \forall \exists^*$; the essentially Monadic Class is a prefix class of formulas with such atoms which after Skolemization will contain no more than one nonconstant term in the set of arguments of the atom. For example, the atom (in Skolemized form) P(a, f(x,y), b, f(x,y)) will be considered to be (essentially) monadic.

The method presented here is a successor to a previous algorithm for the same class developed by the author in [Tam91]. The basis of a new method is completely different from the one in [Tam 91], compared to which the new method is significantly faster for realization, and the correctness/completeness proofs are much simpler.

Our method is practically in no way related to the methods of building finite models described in [Wos 84]; the last methods rely totally on the human user; the theorem-prover (which does not use decision strategies) offers only a helpful housekeeping environment to keep track of the investigated search space.

The proof of the termination of our algorithm uses only the theorems about the completeness of the apriori $>_k$-ordering-refinement of resolution method presented in chapter 6 and the theorem about the completeness of narrowing given in [Sla74], thus being also the proof of the finite controllability of AM.

We have not set ourselve the task to improve the known upper bounds on the size of models for formulas in the Ackermann and Monadic class (see [DG79]). Neither can we show that our method has a better upper bound concerning the computational complexity of building a model than the one given in [Lew 80].

Nevertheless, our algorithm is capable of computing for "small formulas" considerably smaller models than the classical methods given, for example, in [DG79] (such optimisations obviously were not the aim of the authors of [DG79]); it uses more of the structure of the investigated formula, being more flexible, so to say. Since it relies on a certain resolution refinement ($>_k$-refinement), we have the opinion that it shares some of the good properties of resolution. For example, our method performs the satisfiability checking incrementally (satisfiability checking must be performed many times for generating even the smallest models) using a very efficient resolution refinement.

It would be interesting to compare the implementation of our method with the implementation of classical methods; unfortunately no implementations of the classical methods are known to us.

Last not least, we hope that the general outline of the soundness and termination proof of the algorithm might be useful for showing finite controllability and devising analogous algorithms for other solvable classes for which we have the deciding resolution refinements (Classes K and E, for example) and giving some additional insight to related problems.

7.1 THE CLASS AM

The class AM is essentially the class of clause sets corresponding to the Initially extended Ackermann Class and the Essentially Monadic Class, with arbitrary constant substitutions allowed.

<u>DEFINITION</u> 7.1.: Let F be a prenex form of a formula of the first order predicate calculus (without functional symbols and without the equality predicate). Let S_F be the clause form of F. Let $A = P(t_1,...,t_n)$ be an atom in S_F. Remember that by args(A) we denote the set $\{t_1,...,t_n\}$.

Then some clause set S belongs to the Class AM iff there is such a clause set S_1 which is the clause form of some formula in prenex form and for clause C in S can be split into clauses $C_1, \dots, C_n$ so that for each C_i $(1 \leq i \leq n)$ one of the following cases is true:

- (i) C belongs to S_F and for each atom atom A in C at least one of the following two possibilities holds:
  - Args(A) contains at most one non-ground term.
  - Every functional term in Args(A) is of arity ≤ 1 and A contains only one variable (this variable may have several occurences in A).
- (ii) $C = D\sigma$, where D is a clause obeying case (i) above, and $\sigma = \{c_1/x_1, ..., c_n/x_n\}$ for some set of constants $c_1,\dots,c_n$ and some set of variables $x_1,\dots,x_n$.

Obviously the clause form of any formula from either the Initially-Extended Ackermann Class or the Essentially Monadic Class belongs to Class AM. More so, clause forms of "mixes" of these classes also belong to AM. Also notice that AM corresponds to a certain subset of Maslov's Class K.

<u>EXAMPLE</u> 7.1.: The following clause set belongs to the Class AM:

- $\{P(f(x), a, g(x)),\ \neg S(h(x, y, z), a, h(x,y,z)\}$,
- $\{R(f(a),a,b),\ S(h(a,y,z), b, h(a,y,z))\}$,
- $\{S(f(x),x,x),\ \neg P(k(x,y), k(x,y), b),\ P(f(x), g(x), g(x))\}$,
- $\{S(f(x),x,x),\ P(f(y),y,f(y))\}$
- $\{R(f(a), a, g(a))\}$,
- $\{P(x, x, x)\}$,
- $\{P(a,b,c)\}$.

The following clauses and atoms cannot occur in any clause set belonging to AM, as they cannot be obtained by converting a function-free prenex formula to clause form and then by splitting:

- $\{P(f(x), x),\ S(g(y, z))\}$,
- $P(f(x), g(y))$,
- $P(f(x), f(a))$.

<u>DEFINITION</u> 7.2.: In the following we will say that an atom A in the clause set of Class AM is of the Ackermann type if and only if A has at least two different nonconstant arguments; if A is not of the Ackermann type, we will say it is of the Monadic type. Notice that our usage of the names "Ackermann" and "Monadic" does not exactly correspond to the "normal" usage of these names (in our usage Monadic-type atoms can't be also of the Ackermann type). For literals the same classification is defined analogously.

EXAMPLE 7.2.: The following literals in the previous example are of the Ackermann type:

P(f(x), a, g(x)), S(f(x), x, x), P(f(x), g(x), g(x)), R(f(a), a, g(a)).

The following literals in the previous example are of the Monadic type:

¬S(h(x,y,z), a, h(x,y,z)), S(h(a,y,z), b, h(a,y,z)), ¬P(k(x,y), k(x,y), a),

R(f(a), a, b), P(x, x, x), P(a, b, c).

Notice that ground instances of the Ackermann-type literals can be Monadic; that happens iff the Ackermann-type literal contains only one function symbol.

7.2 THE ALGORITHM FOR BUILDING FINITE MODELS FOR FORMULAS IN AM

Recall the $>_k$ -ordering defined in chapter 6 for deciding Maslov's Class K. Recall that by Robinson's resolution method we denote the resolution method where factoring is performed at the time of a binary resolution step only and one of the factorized literals must be the one which is resolved upon.

Throughout the current chapter we will use the refinement of resolution method presented by N. Zamov in chapter 6. We will call it k-refinement:

DEFINITION 7.3.: By k-refinement we will denote the apriori refinement of Robinson's resolution method using the following ordering $>_k$: $A >_k B$ iff A strictly dominates B (for the definition of the dominating relation see chapter 6). Let S be any set of clauses. By $R^*_{>_k}(S)$ we denote the set of clauses derivable from S by the k-refinement.

DEFINITION 7.4.: We will say that some set of clauses G is in the stable form iff $G = R^*_{>_k}(G)$.

Let $S = \{C_1, \ldots, C_n\}$ be a satisfiable clause set from Class AM.

We will use the Herbrand universe $H_S = \{h_1, h_2, \ldots\}$ of S (compare Def. 2.28) as a domain of the model M for S (Herbrand's theorem implies that if a first order predicate calculus formula has a model, it has a Herbrand model - i.e.

a model with the domain of elements of Herbrand universe). For nontrivial formulas the Herbrand universe is infinite, and in order to build a finite model, we must build a finite domain and find an interpretation of the functions of the formula on this domain. Once the finite domain and interpretation of function symbols is found, finding the interpretation of predicate symbols is a relatively easy task, and we won't deal with it in the current presentation.

Whenever we speak of " a model ", it is assumed that interpretations of predicate symbols may be left undetermined in it - it must only determine a finite domain and an interpretation of function symbols on that domain. Whenever we speak of a process of " model building", it is assumed that finding intepretations of predicate symbols may be left out of the process. When we speak of " the size of a model", we will mean the cardinality of the domain.

We will select a set of equations $E = \{e_1, \ldots, e_m\}$ (where each e_i is of the form $h_k = h_l$ for some k, l), that would decompose the Herbrand universe H into a finite number of different equivalence classes $\{c_1, \ldots, c_r\}$ such that the "formula" $SE = \{C_1, \ldots, C_n, e_1, \ldots, e_m\}$ would be satisfiable. In this case SE would have a finite model M with the domain D constructed by choosing one arbitrary element from each c_i and the Skolem functions interpreted by the set E. Obviously M would be a model for S.

In case we knew that any satisfiable formula in AM has a finite model, we could in principle use the following method for finding a finite model: for each possible cardinality of the model (1, 2, ...) check all possible classes of isomorphism of sets of equations, giving the domain of that size from the Herbrand universe. As we assumed that a satisfiable formula in AM always has a finite model, we will eventually find it. For example, suppose we have a clause set in AM with two one-place function symbols, f and g, and one constant symbol a. For cardinality 1 we must check the single set of equations: f(a) = a, g(a) = a. If this does not satisfy our clause set, we have to check the following eight sets for the cardinality two:

$$\{\{f(a) = a, \quad f(g(a)) = a, \quad g(g(a)) = a\},$$
$$\{f(a) = a, \quad f(g(a)) = a, \quad g(g(a)) = g(a)\},$$
$$\{f(a) = a, \quad f(g(a)) = g(a), \quad g(g(a)) = a\},$$

$\{f(a) = a, \quad f(g(a)) = g(a), \quad g(g(a)) = g(a)\},$

$\{g(a) = a, \quad g(f(a)) = a, \quad f(f(a)) = a\},$

$\{g(a) = a, \quad g(f(a)) = a, \quad f(f(a)) = f(a)\},$

$\{g(a) = a, \quad g(f(a)) = f(a), \quad f(f(a)) = a\},$

$\{g(a) = a, \quad g(f(a)) = f(a), \quad f(f(a)) = f(a)\}\}.$

Checking any such set of equations can be performed by substituting a finite set of domain elements given by the equations to all variables in the clauses in all possible ways, and computing the values of functions given by the equations. This gives a large set of propositional clauses, the satisfiability of which can be decided in standard ways (using resolution, for example).

The amount of possible sets of equations grows superexponentially on the cardinality of the model checked, and the exhaustive search method is practically applicable only for searching for very small models. Using exhaustive search (total backtracing for checking all possible interpretations for function symbols) for searching a finite model would prevent finding any but the smallest models: for example, a single functional symbol with arity 2 on a ten-element domain yields 10^{100} different interpretations (as there are ten possible values of this function for each of the 100 argument pairs).

The search algorithm we are going to present does not rely on such kind of an exhaustive search. Instead it will have the property that whenever any equation $t = d$ is considered to be acceptable (that is, the set $\{S \cup E \cup t = d\}$ is satisfiable, where S is our original clause set, E is the set of already accepted equations, and some additional special conditions for $t = d$ are also fulfilled) this equation will be retained throughout the whole search. That is, the set of accepted equations grows monotonously and we never throw away any accepted equation (we do not "backtrace"). Thus at no moment during the model construction process (before the very last, the actual finding of a finite model) do we have a finite domain at hand, and for satisfiability checking we cannot use the conversion to a big propositional formula, as can be done for the exhaustive search described earlier. Therefore we will use the k-refinement of resolution (used in [Zam 89, 89a] and chapter 6) combined with Slagle's narrowing method from [Sla 74], which will yield a decision procedure for checking acceptability of new equations. We will prove that such a non-exhaustive algorithm always terminates - when certain conditions of accepting an equation are imposed. This also gives a proof that any satisfiable formula in AM indeed has a finite model.

Let H be the Herbrand universe of a satisfiable S in AM. We will use the same main idea for building the finite model for S as is used for building finite models in [DG79], for example: we will traverse the Herbrand tree (compare def. 2.31) moving from level 1 to deeper levels and trying to add equations between terms on the current level and some earlier level. Each equation cuts short some branches, until the branches in H are pruned.

The minimal consistent criteria for accepting some equation $t = d$ between Herbrand terms is the satisfiability of a union $\{S \cup E \cup t = d\}$ where E is the set of equations found so far. Such a criteria gurantees that S will have a model with the constructed interpretation of function symbols (but it generally does not guarantee finding a finite model, since we do not backtrack to try all possible combinations of equations). We will give a simple example of the process.

<u>EXAMPLE</u> 7.3.: Clause set $S = \{\{P(x,f(x))\}, \{\neg P(f(y),y)\}\}$ has a Herbrand tree with a single branch: $a \rightarrow f(a) \rightarrow f(f(a)) \rightarrow f(f(f(a))) \rightarrow f(f(f(f(a))))$, ... At first try to add the equation $f(a) = a$. But $\{S \cup f(a) = a\}$ is unsatisfiable, therefore this equation cannot be added. At the next step try to add $f(f(a)) = a$, but again $\{S \cup f(f(a)) = a\}$ is unsatisfiable. Then trying $f(f(a)) = f(a)$ one will fail again. The first equation which can be added consistently is $f(f(f(a))) = a$, which prunes our single branch, thus giving a finite domain with four elements.

In general we have more than one branch and therefore need more than one equation.

One of the most important properties of our model-building algorithm is the absence of backtracing. Unfortunately, as we do not use backtracing, for testing the acceptability of an equation it is not possible to rely just on testing the satisfiability of the union of our original clause set, already obtained equations and the tested equation: we have no means for a proof that such a process of finding the equations will terminate. In order to guarantee termination of the search for a finite domain we will use somewhat stronger criteria. The criteria we have chosen differ from those given in [DG79] for the Ackermann Class, for example, in such a way that as one of the consequences we can almost always build much smaller models for small formulas, and need much less search steps (the number of steps of checking acceptability of single equations are much smaller).

We won't, however, try to improve the known upper bounds on the size of the finite domains.

7.3 THE CRITERIA FOR ACCEPTING A SET OF EQUATIONS

In general we have more than one branch in the Herbrand tree and therefore need many equations to prune all the branches.

In order to guarantee termination of the model-building process without backtracking, we will use a special transformation of the stable form of the input clause set to a certain Essentially Monadic clause set.

During the whole process of search we will use the transformed clause set combined with a special disunification criterion (in respect to the stable form of the original clause set) for checking any new equation t = d.

- The minimal (correctness) criterion: $\{S' \cup E \cup t = d\}$ must be satisfiable, where S' is the transformation of the stable form of the investigated clause set to the Essentially Monadic form and E is the set of equations obtained so far.

- The disunification criterion: if the arity of the function symbol of the left-side term t of the equation t = d is equal to one, then if some pair of atoms A, B in the stable form of the original clause set S is not unifiable, and either A and B have dual signs in S or belong to the same clause in S, then the pair A, B must not be unifiable in the equational theory $\{E \cup t = d\}$.

The case when the arity of the function symbol of the left-side term is bigger than one is rather important. The harmlessness of this "exception" stems from the fact that for the Monadic Class the model building process will terminate without using any special disunification criterion. On the other hand, if this exception is not used, it becomes very hard to show the termination of such a model building process (if it is terminating at all).

EXAMPLE 7.4.: (disunification criterion): Suppose our input set contains Ackermann-type literals $P(f(x), x)$ and $\neg P(y, f(y))$. These literals are not unifiable. However, given an equation $f(a) = a$, the substitution instances $P(f(a), a)$ and $P(a, f(a))$ become unifiable, as then $P(f(a), a) = P(a, a)$ and $P(a, f(a)) = P(a, a)$.

Checking the correctness criterion for any new equation $t = d$ involves deciding the satisfiability of a clause set which contains the equality predicate. The simplest way to deal with the equality would be to axiomatize it in a standard way, giving axioms for transitivity, commutativity, reflexivity, and axioms for substituting into predicates and functions of the formula. Another standard way to handle equality in resolution context would be using paramodulation (see [Wos 84]). Unfortunately, neither the axiomatization (even the transitivity axiom alone) nor paramodulation can be handled by any decision strategy known to the author. Thus we have to use some other ways for handling the equality predicate.

In the following we will use the narrowing method of Slagle (see [Sla 74]): if E is a confluent term rewriting system, narrowing of any expression e is obtained by unifying a left side term h of some rewrite rule in E with some non-variable subterm t of e, applying the obtained substitution $\sigma = \text{m.g.u.}(h, t)$ to e, and rewriting the expression $e\sigma$ to its normal form modulo E (see, for example, [Bar 81]). A narrowing of the narrowing of e is also considered to be a narrowing of e.
A well-known theorem of Slagle [Sla 74] states that if E is a complete term rewriting system (set of simplifiers in his terminology) and S is a set of clauses without equality, then $\{E \cup S\}$ is satisfiable if and only if the full narrowing of S using E is satisfiable. The full narrowing of S is defined as S augmented by all narrowings of all clauses in S.

Remark 1 in forthcoming section 8.4 notes that the set of equations generated by our algorithm can always be oriented to a confluent term rewriting system. It is easy to see that a full narrowing of any set of clauses in AM with any set of equations generated by the the model construction algorithm presented later will also be in the class AM. Thus we have a method for checking the acceptability of sets of equations for formulas in AM: perform full narrowing and then use k-refinement of resolution.

EXAMPLE 7.5.: Let E = {f(a) → a, g(b) → a}. The single narrowing of the atom P(f(x), x) is P(a, a). The atom P(g(x)) has a single narrowing P(a). The atom Q(g(x), f(x)) has two narrowings, Q(g(a), a) and Q(a, f(b)). Consider the set of clauses S = {{P(f(x), x), S(g(x)), {Q(g(x)), ¬S(f(x)), {¬P(f(x), f(x)), P(x,x).

The full narrowing of S using E is the following set:

S ∪
{{P(a, a), S(g(a))},
{P(f(b),b), S(a)},
{Q(g(a)), ¬S(a)},
{Q(a), ¬S(f(b))},
{¬ P(a,a), P(a, a)}}.

As can be easily checked by application of the k-refinement, this full narrowing is satisfiable.

DEFINITION 7.5.: We say that some set of reductions R preserves disunification in S iff the following holds: if two atoms A and B in the set S are not unifiable, then no narrowing A' of A (using E) is unifiable with B or with any narrowing B' of B.

In the following algorithm the disunification preservation is checked by the procedure ok- disunific.

EXAMPLE 7.6.: Take the clause set
S = {{P(x,f(x))}, {¬P(f(y),y)}, {¬P(x,g(x))}, {Q(z), Q(g(z))}, {¬Q(u), ¬Q(g(u))}.

Recall that the Herbrand tree for S

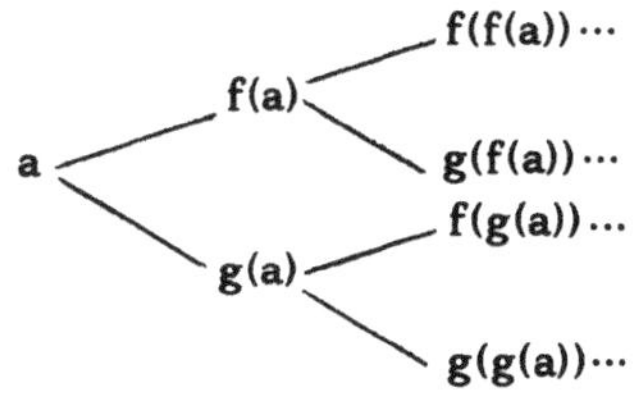

is ordered in the following way: $\langle a, f(a), g(a), f(f(a)), \ldots \rangle$.

We will see how the disunification condition, given S, will treat several reductions.

- $f(a)$ a is not OK, as narrowing gives $P(a,a)$ and $\neg P(a,a)$ from literals $P(x,f(x))$ and $\neg P(f(x),x)$, respectively.
- $f(f(a)) \rightarrow a$ is not OK, as narrowing gives $P(f(a),a)$ and $\neg P(a,f(a))$, accordingly, and $P(f(a),a)$ is unifiable with $\neg P(f(x),x)$.
- $f(f(f(a))) \rightarrow a$ is OK: narrowing gives literals $P(f(f(a)),a)$ and $\neg P(a,f(f(a))$, which are neither unifiable with each other nor with $P(x,f(x))$ and $\neg P(f(x),x)$, respectively.
- $g(a) \rightarrow a$ is not OK, as narrowing gives a clause $\{Q(a)\}$, from the clause $\{Q(z), Q(g(z))\}$, where $Q(z)$ and $Q(g(z))$ are not unifiable.
- $g(g(a)) \rightarrow a$ is OK: narrowing gives

$$\{\neg P(g(a),a)\},\ \{Q(g(a)), Q(a)\} \text{ and } \{\neg Q(g(a)), \neg Q(a)\}.$$

- Now suppose we have already taken $f(f(f(a))) \rightarrow a$ into the set of accepted equations. Thus we have the narrowings $P(f(f(a)),a)$ and $\neg P(a,f(f(a)))$ from $P(x,f(x))$ and $\neg P(f(y),y)$. Consider the reduction $g(f(f(a))) \rightarrow a$: it gives the narrowing $\neg P(f(f(a)),a)$ from $\neg P(x,g(x))$, thus not preserving disunification with respect to the set E (as $P(x,f(x))$ and $\neg P(x,g(x))$ are not unifiable).

7. 4 THE ALGORITHM FOR BUILDING FINITE MODES: DETAILS

We present an algorithm for building a finite domain and an interpretation of the function symbols for a clause set S in Class AM. The full model for S (i.e. also an interpretation of predicate symbols on elements of domain) can then be built from the domain and the interpretation of function symbols in different, simple ways (for example, applying unrestricted resolution - with elimination of subsumed clauses - to the set obtained from S by substituting elements of a domain to variables in S in all possible ways and normalizing all atoms) which are not investigated in our book.

The following algorithm actually constructs a finite domain and interpretation for a clause set $R^{\bullet}_{\succ_k}(S)$. Notice that since $R^{\bullet}_{\succ_k}(S)$ is a set of clauses derived from S, $R^{\bullet}_{\succ_k}(S) - S$ might subsume several clauses in S and several function and predicate symbols in S might not appear in $R^{\bullet}_{\succ_k}(S) - S$. In the model of S these symbols can be interpreted in an arbitrary way.

7. 4. 1 The model construction process

Let S be the investigated satisfiable set of clauses. At first the stable form $R^{\bullet}_{\succ_k}(S)$ for S is computed . $R^{\bullet}_{\succ_k}(S)$ is then transformed to a certain Essentially Monadic clause set SNEW, which will be used during the whole model construction process instead of S or $R^{\bullet}_{\succ_k}(S)$.

Consider the ordering of the Herbrand tree in the top-down, left-to-right direction. The previous example: given two one-place function symbols f and g and a single constant symbol a, the first three levels of the tree can be ordered in the following way: ⟨a, f(a), g(a), f(f(a)), g(f(a)), f(g(a)), g(g(a))⟩. For the k-th element h_k of such an enumeration the algorithm will try to add an equation $h_k = h_l$; $(l < k)$, starting with $l = 1$ and continuing up to $l = k_1$, unless the acceptable equation is found earlier. The check for acceptability of any equation $h_k = h_i$ contains following two steps:

(1) in case the arity of the function symbol of the term h_k is equal to one, check disunification preservation,

(2) check satisfiability of {SNEW $\cup$ E $\cup$ h_k = h_i} . The satisfiability check is performed by computing a full narrowing of SNEW with {E $\cup$ $h_k \rightarrow h_i$ }. Actually only the narrowing of the formula SNEW' with the reduction $h_k \rightarrow h_i$ has to be computed, where SNEW' is the narrowing of SNEW using E, obtained ealier. In case the acceptable equation h_k = h_i has been found for some i, one "branch" in the Herbrand tree has been cut. If such an equation is not found, the term h_k is taken as a new element of the constructed domain. In order to speed up future computations, it is useful to introduce a new constant symbol c_k for the term h_k and to compute the stable form of SNEW' with the defining reduction $h_k \rightarrow c_k$. In the following this new stable form will be used instead of SNEW' (obviously this stable form is always satisfiable). One of the important consequences of replacing a possibly deep ground term with a new constant is that the narrowed clause set is always in the Class AM - respectively in the Essentially Monadic subset of AM .

In the following algorithm the satisfiability check is performed by the function satisfiable, which is implemented by the k-refinement. The disunification check is performed by the function ok-disunific okdisunific (S^*, E $\cup$ {fa $\rightarrow$ f b}) could be computed in the following way: for all pairs of nonunifiable dual literals in S^* generate all narrowings of both elements of the pair with {E $\cup$ {fa $\rightarrow$ fb}} and test whether any narrowing of the first element (or the first element itself) doesn't unify with any narrowing of the second element (or with the second element itself). In fact, one should perform this narrowing-computing incrementally during the process of finding new elements of E.

In implementing the following algorithm one should avoid the duplicate calculation of $R^*_{\succ_k}$(narrow(SNEW, fa $\rightarrow$ fb)) by using the value computed earlier during a computation of satisfiable(narrow(SNEW, fa $\rightarrow$ fb)).

```
{COMMENT: S is the satisfiable input set of clauses};
S* := R*>k(S); SNEW :=transform—to—monadic(S*);
D:= {c | c is a constant symbol in SNEW};
IF D = {} THEN D := {a};
DNEW :=D; DLAST :=D; E = {};
WHILE DNEW == {}
 BEGIN
  DNEW := {};
  FOR EACH f IN {h | h is a functional symbol in SNEW }
   BEGIN
    FOR EACH fa IN {f(t1,...,tn) | ∀1≤i≤n: ti∈D and ∃j: tj∈DLAST}
      BEGIN
        foundflag :=FALSE;
        FOR EACH c in D WHILE foundflag = FALSE
          BEGIN
             IF ((arity(f) > 1) OR
                (ok disunific (S*, E ∪ {fa → c})))
                AND
                satisfiable (narrow(SNEW, {fa → c}))
             THEN
                BEGIN
                   E:=E ∪ {fa → c};
                   SNEW := R*>k(narrow(SNEW, {fa → c}));
                   foundflag :=TRUE;
                END
          END
        IF foundflag =FALSE
          BEGIN
             c:=newconstant ();
             E:=E ∪ {fa → c};
             DNEW:=DNEW ∪ {c};
            SNEW := R*>k (narrow(SNEW, {fa → c}));
          END
      END
   END
  DLAST := DNEW; D:= D ∪ DNEW
END.
```

Remark 1:

The set E of all equations generated at each step of our algorithm is always such that orienting all equations $h_k = h_l$ in E into rewrite rules $h_k \rightarrow h_l$ produces a complete term rewriting system.

In view of remark 1, in the following we will treat E as a set of rewrite rules or as a set of equality units, depending on the context.

7.5 TRANSFORM-TO-MONADIC PROCEDURE

In the following we will present a transformation of a stable form S of clause set in the Class AM to a clause set S' in the Essentially Monadic Class so that the following two assertions hold:

- (1) if S is satisfiable, S' is also satisfiable;

- (2) for any set of equations E computed by our finite-model-building algorithm, if $\{E \cup S'\}$ is satisfiable and E satisfies the disunification condition for S, then $\{E \cup S\}$ is satisfiable, too.

The inverse of the assertion (2) is not generally true, however: satisfying the disunification condition and S' will not imply satisfying S. This is the reason why we have to use the transformation in the actual model-building process - if the computed set of equations (satisfying the disunification condition) satisfies S iff it satisfies S', the transformation could be used only in the termination proof and not in the actual algorithm.

At first we will present a pseudo-code version of the transformation, to be followed by a natural-language description and examples.

```
BEGIN
Let us initially have S={C_1,..., C_n} such that R•_>k(S)=S;
S_1 :={};
MON := {R | R is a monadic-type literal in some C_i};
FOR EACH CLAUSE C_i IN S DO
  BEGIN
    S_1:= S_1 ∪ C_i ;
    ACK:={L | L is an Ackermann-type literal C_i};
    FOR EACH LITERAL L IN ACK  DO
      BEGIN
        FOR EACH R IN MON  DO
          BEGIN
            IF EXISTS m.g.u(L, R^d)
             THEN
              BEGIN
               σ:=m.g.u.(L, R^d);
               S_1:= S_1 ∪ C_i σ ;
              END
          END
      END
  END
ACK := {L | L is an Ackermann-type literal in some C_i}
FOR EACH L_i IN ACK  DO
  BEGIN
    sgn:=sign(L_i);
    p[i]:=new-monadic-predicate-symbol();
    {COMMENT: p[] is an array of monadic predicate symbols};
    F(L_i):= {f | f is a function symbol in L_i;
    {COMMENT: let F(L_i), now have a form f_1, ... ,f_m }}
    t:= argument of a functional term in L_i;
    {COMMENT: for Ackermann-type literals t is unique};
```

```
newmon1[i]:= {sgn (p[i](f_1 (t))), ..., sgn(p[i](f_m (t)))}
                      for each f_j  in F(L_i);
   {COMMENT: newmon1[] is an array of sets of literals};
 END
FOR EACH L_i in ACK  DO
 BEGIN
   newmon2[i]:=newmon1[i];
   {COMMENT: newmon2[] is an array of sets of literals};
   FOR EACH R_j IN ACK  SUCH THAT i ≠ j DO
     BEGIN
       IF EXISTS m.g.u(L_i, R_j^d);
        THEN
        BEGIN
          σ := m.g.u(L_i, R_j^d);
          IF NOT(ground(L_i σ ))
           THEN σ:= {y/x},
               where x is a variable in R_j,
               and y is a variable in L_j;
          newmon2[i] := newmon2[i] ∪ (newmon1[j]σ);
          {COMMENT: newmon1[j]σ  ={e_1 σ, ... ,e_N σ}
        END
     END
 END
```

To form the set of clauses S' from S_1, replace each clause C in S_1 by a set of new clauses obtained by replacing all Ackermann-type literals L_i in C with the atoms from corresponding newmon2[i] in all possible ways. During that replacement, whenever a non-ground literal L_i with a single variable x in C is replaced by a ground unary atom L', where a constant symbol c is the argument of some functional term, apply the substitution c/x to the result of transformation.

From the resulting set of transformations remove all such transformed clauses, which contain at least two such transformed literals $A(f_i(c_j))$, $B(f_k(c_l))$, where the corresponding original literals A(...) and B(...) both contained variables and the substituted constants c_j and c_l are different.

END.

In the following we will present the same transformation algorithm in natural language, combined with examples.

Let S be the stable form of a set of clauses in the Class AM (that is, S = $R^{*}_{>k}(S)$). Form the new set of clauses S' = transform–to–monadic(S) by replacing each old clause C from S by a set of new clauses in the following two-step way:

First step:

Form the new set S_1 by replacing each clause C in S by a set of new clauses in the following way:

(1) C contains Ackermann-type literals which are unifiable with some dual Monadic-type literals in S: Let L be such an Ackermann-type literal in C, and let G be some Monadic-type literal in S, the dual of which is unifiable with L. Then m.g.u(L, G^d) will assign a certain constant symbol c for a single variable x in L. Let Cset = $\{c_1, \dots, c_n\}$ be the set of all constants assigned to x by unification with different Monadic-type literals. Form the set of clauses C, $\{C\sigma_1, \dots, C\sigma_n\}$, where $\sigma_i = c_i/x$, where c_i belongs to Cset, x is the single variable in Ackermann-type literals in C.

<u>EXAMPLE</u> 7.7.: Let us transform a clause {P(f(x), x), S(x, c)}, and let the set S contain Monadic literals ¬P(y, a) and ¬P(z, b). Then, as an Ackermann-type literal P(f(x), x) is unifiable with the duals of ¬P(y, a) and ¬P(z, b), we have to generate the set { {P(f(x), x), S(x, c)}, { P(f(a), a), S(a, c)}, { P(f(b), b), S(b, c)}}. Notice that the instances of the Acker-mann-type atoms in the new clauses generated this way are always of the Monadic type.

(2) C does not contain Ackermann-type atoms, unifiable with negation of Monadic-type atoms in S: Let the new set of clauses contain the unchanged clause C alone; that is, in this case additional clauses are not generated.

Thus S_1 will be equal to S ∪ {SINST}, where SINST is a certain set of instances of clauses from S.

EXAMPLE 7. 8.: Consider the clause set S:

{{R(f(x), g(x), x), ¬P(x, f(x))},
{P(a, y)},
{R(f(a), g(a), a)},
{S(f(x), x, x), L(x)},
{¬S(f(x), x, a)},
{¬S(f(x), x, b)},
{L(a)},
{L(b)}}.

As there is a single pair ¬P(x, f(x)), P(a, y) of unifiable dual Ackermann-Monadic literals in S, the first step of the transformation will give us a following set S_1:

{{R(f(x), g(x), x), ¬P(x, f(x))},
{R(f(a), g(a), a), ¬P(a, f(a))},
{P(a, y)},
{R(f(a), g(a), a)},
{S(f(x), x, x), L(x)},
{¬ S(f(x), x, a)},
{¬ S(f(x), x, b)},
{L(a)},
{L(b)}}.

Second step:

Let ACK be the set of all Ackermann-type atoms in S_1. For each atom in ACK take a new unique unary predicate symbol. Also, for each literal L in ACK find the set of all dual literals in ACK, which are unifiable with L. For each element R in such a set one of the following holds: the literal R is syntactically equal to L up to renaming variables, or the unification of R with L produces a substitution σ so that Lσ is ground. Take the set of all function symbols in L - let the set be denoted by F(L). Form a new set of atoms newmon2[i], where i is the number of L in some enumeration of ACK:

For each new monadic predicate symbol P, corresponding to some Ackermann-type literal R, unifiable with the dual of L, form the set of atoms of the form $P(f_i(t))$, where f_i is some element of F(L) and t is determined in the following way:

- In case R is syntactically equivalent to L (up to renaming variables), let t be the argument of any functional term in L.
- In case R is not syntactically equivalent to L (up to renaming variables), then m.g.u.(R, L^d) = {c/x, c/y}, where x and y are the single variables of R and L, correspondingly, and c is some constant symbol (or m.g.u(R, L^d) = {c/x} or m.g.u.(R, L^d) = {c/y}). Take t to be the constant symbol c.

To form the set of clauses S' from S_1, replace each clause C in S_1 by a set of new clauses obtained by replacing all Ackermann- type literals L_i in C by the atoms from corresponding newmon2[i] in all possible ways. During that replacement, whenever a non-ground literal L_i with a single variable x in C is replaced by a ground unary atom L', where a constant symbol c is the argument of some functional term, apply the substitution {c/x} to the result of the transformation.

From the resulting set of transformations remove all such transformed clauses, which contain at least two such transformed literals $A(f_i(c_j))$ and $B(f_k(c_l))$, where the corresponding original literals A(...) and B(...) both contained variables and the substituted constants c_j and c_l are different.

An example of the last "filtration" step: Let us have a clause {P(f(x), x), R(f(x), x)}, and let the corresponding sets of new atoms for the original atoms in the clause be {P1(f(x)), P2(f(a)), P3(f(b))} and {R1(f(x)), R2(f(a)), R3(f(b))}, respectively. Then the transformations {P2(f(a)), R3(f(b))} and {P3(f(b)), R2(f(a))} must be removed, and the resulting set of transformations will be {P1(f(x)), R1(f(x))}, {P2(f(a)), R2(f(a))}, {P3(f(b)), R3(f(b))}.

<u>EXAMPLE</u> 7. 9.: We continue the example from the "first step": The set ACK for S would be

{R(f(x), g(x),x), P(x, f(x)), R(f(a), g(a), a), S(f(x),x,x), S(f(x),x,a), S(f(x),x,b)}.

Let the set of new predicate symbols for ACK be {R1, P1, R2, S1, S2, S3}. The corresponding sets of newmon2[i], are:

R(f(x),g(x),x) - {R1(f(x)), R1(g(x)), R2(f(a)), R2(g(a))},

P(x,f(x)) - {P1(f(x))},

R(f(a),g(a),a) - {R2(f(a)), R2(g(a)), R1(f(a)), R1(g(a))},

S(f(x),x,x) - {S1(f(x)), S2(f(a)), S3(f(b))},

S(f(x),x,a) - {S2(f(x)), S1(f(a))},

S(f(x),x,b) - {S3(f(x))}, S1(f(b))}.

Here is the resulting S' of the second and final transformation step:

{{R1(f(x)), ¬P1(f(x))},

{R1(g(x)), ¬P1(f(x))},

{R2(f(a)), ¬P1(f(a))},

{R2(g(a)), ¬P1(f(a))},

{R2(f(a)), ¬P(a,f(a))},

{R2(g(a)), ¬P(a,f(a))},

{R1(f(a)), ¬P(a,f(a))},

{R1(g(a)), ¬P(a,f(a))},

{P(a,x)},

{R2(f(a))},

{R2(g(a))},

{R1(f(a))},

{R1(g(a))},

{S1(f(x)), L(x)},

{S2(f(a)), L(a)},

{S3(f(b)), L(b)},

{S2(f(a))},

{S3(f(b))},

{L(a)},

{L(b)}}.

7.6 SOUNDNESS

Let S be a stable clause set in the Class AM, clause set S' be equal to transform–to–monadic(S), and E be a certain set of ground equations, computed at any stage during our finite-model-building algorithm for S. By the definition of our model-construction algorithm E satisfies the disunification condition for S.

We will say that a clause C' corresponds to the clause C if C' is a transformed version of C, or the derivation of C can be rebuilt into a derivation of C' by replacing clauses in the original derivation with transformations, if needed. We will use the notion of correspondence also in the inverse sense.

THEOREM 7. 1.: If S is satisfiable then, S' is satisfiable.

Proof:

Recall that during the transformation of S to S' ground substitutions could be applied during the derivation of a transformed clause from the original one. For such new clauses the k-refinement may allow more literals to be resolved upon than for the original clause without ground substitution applied. In order to avoid complications stemming from the k-refinement, we will use the following construction: Build a new set of clauses SG = S $\cup$ G, where

G = {Cσ | (C belongs to S and

C contains a literal L, which contains a single variable x, and

σ = {c/x}, where c is some constant in S)}.

Obviously S is satisfiable iff SG is satisfiable.

We will show that for any clause C' derivable from S' by $R_{>k}$ (without using subsumption) there is a corresponding clause C derivable from SG by $R_{>k}$ (without using subsumption).

Induction basis: The construction of S' from S.

Induction step: We will give the proof for the binary resolution only, as this proof can be easily modified for factorization.

Let $C_1' = \{L_1', ..., L_n'\}$ and $C_2' = \{G_1', ..., G_m'\}$ be two clauses derivable from S' by $R_{>k}$. Let a clause $C_3' = \{L_2', ... , L_n', G_2', ..., G_m'\}\sigma$, where σ = m.g.u(L_1', $G_1'^{d}$), be derivable from C_1' and C_2' by binary resolution. As an induction hypothesis suppose that there are corresponding clauses $C_1 = \{L_1, ... ,L_n\}$ and $C_2 = \{G_1, ..., G_m\}$ derived from SG by $R_{>k}$. We will show that then the corresponding clause $C_3 = \{L_2, ..., Ln, G_2, ... ,G_m\}\sigma$ or a more general clause $C_3 = \{L_2, ..., L_n, G_2, ..., G_m\}\rho$

where $\sigma = \rho\mu$ for some substitution μ , can be derived from C_1 and C_2. Actually, as the next theorem demonstrates, more general clauses cannot be derived; for the current theorem this fact is not needed.

Consider the following cases:

- L_1 and G_1 are Monadic: Then the corresponding literals L_1' and G_1' are syntactically equal to L_1 and G_1 (as Monadic-type literals are not transformed), and derivability of C_3 from C_1 and C_2 is obvious.
- L_1 and G_1 are of the Ackermann type: By the definition of the transformation algorithm, unification of the transformed literals can only be restricted when compared to the unification of original literals. Thus, as L_1' and G_1' are unifiable with m.g.u σ, L_1 and G_1 are also unifiable either with the same m.g.u. σ or a m.g.u. ρ, more general than σ ($\sigma = \rho\mu$ for some substitution μ).
- L_1 is of the Ackermann type, G_1 is of the Monadic type: Notice that in this case L_1' must be a ground Monadic-type literal by the definition of transformation (during the transformation to L_1' the substitution m.g.u(L_1, G_1) = ρ is applied to L_1, and ρ is for the current case always ground; it has the form $\{c/x, f(c)/y\}$, where x is a single variable in L_1, y is a single variable in G_1 (if G_1 is ground, ρ has the simple form $\{c/x\}$), f is a single function symbol in L_1). Thus, L_1 and G_1^d are obviously unifiable and C_3 is derivable from C_1 and C_2.

Q.E.D.

THEOREM 7.2.: If $\{E \cup S'\}$ is satisfiable then $\{E \cup S\}$ is also satisfiable.

Proof:

We will show that for any clause in the full narrowing SN of S using E and for any clause derivable from SN by $R_{>k}$, there is a corresponding clause derivable from the full narrowing S'N of S' by E using $R_{>k}$. Notice that for any non-transformed clause the k-refinement ordering of that clause either exactly corresponds to or is stronger than k-ordering of the corresponding transformed clause.

Induction basis: The full narrowing of S by E. Due to the fact that for any equation $t \rightarrow d$ in E there is a single possible narrowing of any Ackermann-type atom in S by this equation, for any narrowable clause in S by E there is a corresponding clause in S' which can be narrowed in the same way.

Induction step: We will give the proof for the binary resolution only, as this proof can be easily modified for factorization.

Let $C_1 = \{L_1, ..., L_n\}$ and $C_2 = \{G_1, ..., G_m\}$ be two clauses derivable from SN by $R_{>k}$. Let a clause $C_3 = \{L_2, ..., L_n, G_2, ..., G_m\}\sigma$, where σ = m.g.u.$(L_1, G_1{}^d)$, be derivable from C_1 and C_2 by binary resolution. As an induction hypotheses suppose that there are corresponding clauses $C_1' = \{L_1', ..., L_n'\}$ and $C_2' = \{G_1', ..., G_m'\}$ derived from S'N by $R_{<k}$. We will show that then the corresponding clause $C_3' = \{L_2', ..., L_n', G_2', ... , G_m'\}\sigma$ can be derived from C_1' and C_2'.

Consider the following cases:

(1) L_1 and G_1 are Monadic: Then the corresponding literals L_1' and G_1' are syntactically equal to L_1 and G_1, and derivability of C_3' from C_1' and C_2' is obvious.

(2) L_1 and G_1 are of the Ackermann type: As they are unifiable, and the set E preserves disunification, then there must be such transformations of C_1 and C_2 which have literals L_1' and G_1' corresponding to L_1 and G_1, so that L_1' and G_1' are unifiable (notice that this is generally true only due to disunification preservation by E). We have to investigate m.g.u(L_1, G^d_1):

(2.1) m.g.u.(L_1, G^d_1) = {x/y}, where x is a single variable in L_1, y is a single variable in G_1: There are such corresponding C_1' and C_2' that m.g.u(L_1', G^d_1) = {x/y} for corresponding x, y.

(2.2) m.g.u.(L_1, G^d_1) = {c/x}, where x is a single variable in L_1:
Consider the "leaf" clause S (contained in S'N) in the derivation of C_1 containing the literal L_1. There is a transformation S' of S in S'N with a substitution {c/x} applied. The derivation of C_1 can be easily rebuilt into a derivation of C_1' from S'N, replacing corresponding literals and applying the substitution {c/x} where needed.
Now we have got a half of C_3', namely $\{L_2', ..., LN'\}\sigma$, with a right substitution. The other half, $\{G_2', ... , GM'\}\sigma$ is obtained by applying a substitution m.g.u$(L_1', G_1'^d)$ = {c/y}.

(2.3) m.g.u.(L_1, G_1^d) = {c/x, c/y}, where c is some constant,

x is a single variable in L_1, y is a single variable in G_1: This case is similar to the previous case, with the main difference that the argumentation from the last case could be carried out symmetrically. We can freely pick one of the possible two ways of analysis:

Consider the "leaf" clause (contained in SN) S in the derivation of C_1 containing the literal L_1. There is a transformation S' of S in S'N with a substitution {c/x} applied. The derivation of C_1 can be easily rebuilt into a derivation of C_1' from S'N, replacing corresponding literals and applying the substitution {c/x} where needed.
In this way we get a half of C_3', namely $\{L_2', ..., LN'\}\sigma$, with a right substitution. The other half, $\{G_2', ... , GM'\}\sigma$ is obtained by applying a substitution m.g.u.$(L_1', G_1'^d)$ = {c/y}.

(2.4) m.g.u.(L_1, G_1^d) is an empty set. Similarly to case (2.1) L_1 is syntactically equal to G_1 and derivability of C_3' is trivial.

(3) L_1 is of the Ackermann type and G_1 is of the Monadic type (e.g, P(f(x),x) and $\neg$P(x,b)). Obviously the corresponding literal L_1' has a new predicate symbol, different from the one in L_1, and thus L_1' and $G_1'^d$ are not unifiable. But, during the first step of the transformation, for all clauses C containing an Ackermann-type atom L, unifiable with some Monadic-type atom P in S with some σ = m.g.u.(L, P^d), a new Monadic instance $C\sigma$ was generated. Consider the derivation of a clause C_1 from a full narrowing SN. Let σ = m.g.u.(L_1, G_1^d). Clearly the derivation of C_1 can be rebuilt into the derivation of $C_1 \sigma$ by replacing the clause C in SN containing the literal L_1 (as L_1 is of the Ackermann-type and unifiable with a Monadic atom, it must be non-ground) with a clause $C\sigma$, found in S'N, and applying a substitution σ where needed. Therefore C_3' = C_3 can be derived from S'N, like in case (1) of two Monadic literals.

Q.E.D.

7.7 TERMINATION PROOF

Given that our original formula S belongs to Class AM, the termination of every subprocedure in our program is easy to prove.

We have to show that $R^{\bullet}_{>k}$ always terminates. We use the fact that $R_{<k}$ decides Maslov's Class K, thus also the Class AM.

If $R^{\bullet}_{>k}$ terminates on a set S, then procedure *satisfiable* terminates on S also.

The transformation algorithm transform-to-monadic oviously terminates. The same is true for okdisunific.

7.7.1 Termination of the whole model construction algorithm

We prove the termination by giving an upper bound on the length of the branches of the "pruned" Herbrand tree. Recall that the finite domain D is constructed by finding a set of rewrite rules E such that terms containing elements of D are reduced by rewrite rules in E to terms in D.

The idea of the proof is the standard pigeonhole argument, used in [DG 79] for most termination proofs of the kind. In fact, one could "read off" the current termination proof from some proofs in [DG79]. However, we will give a termination proof in order to have a complete presentation.

During the proof we will consider the modification of the model construction algorithm where new constant symbols are not introduced for new domain elements, and possibly deep ground terms are used instead. Obviously there is no difference in the process of model construction between the modification with new constant symbols and the modification without them.

Let $B_n = \langle t_1,\ldots,t_n \rangle$ be a tuple of m first elements of the constructed domain D. Let $B_m = \langle d_1,\ldots,d_i \rangle$ be such a subtuple of B_n, which contains exactly the elements of a certain branch of Herbrand universe. For example, $\langle b, f(b), g(c,f(b)), f(g(c,f(b)) \rangle$

could be such a branch. As our algorithm has added all elements of B_n to D, it also has added all elements of the subtuple (branch) B_m, and therefore cannot have added any equation of the form $d_i = t$ (where t is any term and $1 < i < m$) to E, since adding any such equation to E would have meant pruning the branch B_m. Nevertheless, during the generation of D the algorithm has for every equation of the form $d_i = d_j$ (where d_i and d_j belong to the branch B_m and $1 < i < m$ and $1 < j < i$) examined the possibility to add this equation to E. There are two possible causes for rejecting the examined equation $d_i = d_j$:

(A) equation $d_i = d_j$ does not preserve disunification in S.

(B) $\{SNEW \cup E \cup d_i = d_j\}$ is not satisfiable.

In the following we show that both of the abovementioned conditions, (A) and (B), decompose every branch of the Herbrand tree of S into a finite number (n_A, n_B, accordingly) of classes so that if an element d_i belongs to the same class (for both of the conditions A, B) as an element d_j, then the equation $d_i = d_j$ would be accepted by our algorithm. From this it immediately follows that if the number m (which represents the number of domain elements constructed) $> n_A * n_B$, the set E will contain one equation of the form $d_i = t$ ($1 < i < m$, t is some term), thus pruning the branch B_m in D. The upper bound on the size of the model which will arise from the upper bound on the length of branches is by no means the best: we give it only in order to prove the termination of the search. We say that a term contains itself at depth 0, arguments of its leading functional symbol at depth 1, etc.

By the inequality of structures up to some depth n we mean that when we replace the subterms deeper than n both in d_i and d_j by a new common constant c, the resulting terms are not syntactically equal. For example, the structures of terms f(a, g(h(b,c), c)) and f(a, g(g(d,c),c)) are equal up to depth 1, but inequal up to depth 2.

(A): disunification

Equation $d_i = d_j$ may not preserve disunification for atoms in S only if one of the following holds:

- the leading function symbol of d_i has an arity equal to one and d_i contains d_j as a subterm at depth less than 3: We can say that "the cause" for this condition is the occur check: take atoms P(f(x),x), P(y,f(y)) and equation f(f(f(f(a)))) = f(f(f(a))), for example. The chosen atoms are not unifiable, but their narrowings with the chosen equation are both P(f(f(f(a))),f(f(f(a)))). The depth boundary 3 is sufficient since Ackermann-type atoms in S do not contain more than one variable and depth of terms in these atoms is 0 or 1.

- the structure of d_i is up to depth 1 not equal to the structure of d_j. The number of classes of terms with different structure up to depth 1 is equal to $(n_f * (n_c + n_f)^m$ where n_f is the number of the functional symbols in S, n_c is the number of constant symbols in S, and m is the maximal arity of function symbols in S.

Overall, we can take the number for classes of disunification $n_A = 3 * (n_f * (n_c + n_f)^m$ for the following reasons. Consider the branch B_m. Divide it into parts consisting of three adjacent elements. For example, given a branch ⟨ b, f(b), g(c, f(b)), f(g(c, f(b)), f(f(g(c, f(b))), h(f(f(g(c, f(b))))), ...⟩ we will divide it into ⟨⟨b, f(b), g(c,f(b))⟩, ⟨ f(g(c,f(b)), f(f(g(c,f(b))), h(f(f(g(c,f(b)))))⟩, ... ⟩ . Now for any two such parts of the branch the "occur check" reason for disunification cannot hold for any pair of elements taken from these parts, accordingly. Thus only the structural differences have to be considered for testing the disunification property between elements of any two three-element parts of the branch. In fact, we could organize the search for an equation guaranteeing disunification preservation by investigating a single branch only, and considering only the first elements of three-element parts of the branch.

(B): satisfiability

$n_B = 2^w$, where w is the number of different atoms in the set of clauses transform–to–monadic($R^*_{>k}$(S)). The following explains our selection of the number n_B.

Differently from formulas in the Ackermann Class, the satisfiability of a set of ground equations for any formula in a Monadic Class is equivalent to the following criterion (see [DG79] for the well-known, slightly different version of this criterion):

(1) Form the set of all instantiations of all atoms in a clause set $R^*_{>k}$(S),

obtained by applying substitutions given by unifying the left sides of all equations $t \rightarrow d$ with terms in the atoms. For example, given literals $\{P(f(x,y)),\ R(g(x))\}$ and equations $\{f(a,b) \rightarrow c,\ g(e) \rightarrow 1\}$ we will get a set $\{P(f(a,b)),\ R(g(e))\}$.

(2) Rewrite the terms in the set obtained after the previous step (1) to normal forms given by the set of ground equations. From the previous example we will get a tuple $\{P(c),\ R(1)\}$.

Steps (1) and (2) give us two tuples of atoms with one-to-one correspondence between elements: atoms with certain ground substitutions and corresponding normal forms.

(3) Give all the atoms in the tuples signs (either positive or negative), so that corresponding atoms in two tuples would have the same sign. For example, $\{\neg P(f(a,b)),\ R(g(e))\}$ and $\{\neg P(c),\ R(1)\}$.

Now the investigated set of equations is satisfiable for the investigated clause set iff it is possible to give such a combination of signs in the described way, so that the conjunction of all literals in two signed tuples with the investigated formula is satisfiable.

For the Monadic Class the described procedure gives a well-known upper bound on the size of the finite domain: 2^m, where m is the number of different predicate symbols in the clause set. For the Essentially Monadic Class the upper bound is obtained in an analogous way, and the number of different atoms in the stable form of a formula can be taken as an upper bound for m.

For the Ackermann Class the reasoning above cannot be applied. For example, consider the reduction $f(a) \rightarrow f(b)$ and an atom $P(f(x),x)$. The instance of $P(f(x),x)$ obtained from the left side of the reduction is $P(f(a),a)$, which is reduced to $P(f(b),a)$. The instance $P(f(b),b)$ obtained from the right side of reduction is different from $P(f(b),a)$.

Chapter 8

APPLICATIONS

8.1 KL-ONE type languages

8.1.1 KL-ONE and predicate logic

The knowledge representation system KL-ONE was introduced in [BS 85] and a large number of systems based on the similar ideas have been built, for example BACK [NvL 88], CLASSIC [BBMR 89], KANDOR [Pat 84], KL-TWO [Vil 85], KRYPTON [BPGL 85], LOOM [MB 87], NIKL [KBR86]. Generally speaking, KL-ONE type languages can be viewed as certain frame-based, database-oriented languages, where heavy stress is put on the definition of concepts, which can be viewed as unary predicates. Another common feature of these systems is the separation of the knowledge into a terminological part (definitions of concepts), called "T-box" and an assertional part (database), called "A-box". There is a special concept language for defining concepts in the T-box, which can be translated easily to (first order) predicate logic. Actually the concept language corresponds to a certain fragment of predicate logic.

<u>EXAMPLE</u> 8.1.: Given the concepts "person", "female" and "shy", the concept "persons who are female or not shy (pwafons)" can be expressed by the formula

pwafons = (and person (or female (not shy)))

in concept logic, with the translation

$$(\forall x)\, \text{pwafons}(x) \Leftrightarrow (\text{person}(x) \wedge (\text{female}(x) \vee \neg\, \text{shy}(x)))$$

in predicate logic.

The user of the language can describe concepts in the T-box, write down known facts in the A-box, and ask queries from the system, similarily like it is done in PROLOG. Differently from PROLOG, however, the language is decidable - any query can be answered (either positively or a negatively) in finite time. Several other important properties are also required to be decidable. For example, the question whether a T-box is satisfiable, or the question whether one concept P is a subconcept of another concept Q. The last question is called "subsumption problem" in the KL-ONE context: the problem can be read as "are all objects of type P also of type Q?". For example, given a definition of "female" as a "person" who is not "man", the concept "female" is subsumed by the concept "person". Several KL-ONE type systems have the capability to compute the full tree defined by the relation of concept subsumption, and this capability is considered to be rather important from the practical point of view.

So far no language exists which could be called "the KL-ONE language" - every system implements a language with different restrictions and different extensions. However, the language called ALC which has been introduced in [SS 88] can be considered as a basic language for KL-ONE systems, on which different extensions can be built. In [SS 88] it is shown that satisfiability and KL-ONE-subsumption in ALC are PSPACE-complete.

ALC contains two kinds of basic symbols, denoting concepts (corresponding to unary predicates) and roles (binary predicates). ALC is intended as a media for writing sets of concept definitions. We review a definition of the formulas of ALC, called concept description (compare e.g. [BBHNS90]). Simultanously, we define for every concept description of ALC a corresponding formula F' of predicate logic, called translation of F:

<u>DEFINITION</u> 8.1: Any concept description of ALC has one of the following forms:

- ***TRUE*** - top concept of the subsumption hierarchy. Translation: ***TRUE*** ≡ the truth value "true" (or some tautology).

- ***FALSE*** - bottom concept of the subsumption hierarchy. Translation: ***FALSE*** ≡ the truth value "false" (or some contradiction).

- ***P***, where ***P*** is a concept symbol (representing an atomic concept). Translation: P' ≡ P(x), where P is a predicate symbol uniquely corresponding to P; x, is a free variable.

- ***(not F),*** where ***F*** is an arbitrary concept description of ALC. Translation: $\neg F'$, where F' is the translation of ***F***.

- ***(and F G)***, where ***F*** and ***G*** are arbitrary concept descriptions. Translation: $(F' \wedge G')$, where F' and G' are the translations of ***F*** and ***G***, respecitively.

- ***(or F G)***, where ***F*** and ***G*** are arbitrary concept descriptions. Translation: $(F' \vee G')$, analogously to conjunction.

- ***(all R F),*** where ***R*** is a role, and ***F*** is an arbitrary concept description. Translation: $(\forall y)(R(x,y) \Rightarrow F'[y])$, where x is a (new) free variable and F' contains no free variables except y. (This is what the notation $F[y]$ indicates).

- ***(exists R F)***, where R is a role, and ***F*** is an arbitrary concept description. Translation: $(\exists y)(R(x,y) \wedge F'[y])$, where x is a (new) free variable and F' contains no free variables except y.

<u>DEFINITION</u> 8. 2.: A concept definition in ALC is an expression of the following form: $P = F$, where ***F*** is a concept description and ***P*** a concept symbol. Translation: $(\forall x)(P(x) \Leftrightarrow F'[x])$.

A very important additional restriction to the described language is that no circularities are allowed in concept definitions - that is, there must be a certain hierarchy of concepts such that all concepts in the right side of any concept definition are lower w.r.t. the hierarchy than the defined concept. For example, it is not allowed to define a human as an animal whose mother is human: ***human = (and animal (exists mother human)),*** since this definition contains a circularity.

Now, considering the properties of satisfiability and universal satisfiability of concept descriptions as defined e.g. in [BBHNS90], we may state the following:

<u>Proposition</u> 8.1.: A concept description D is <u>satisfiable</u> iff the first order formula $((\forall x)P(x) \Leftrightarrow D'[x])$ and $(\exists x)P(x)$ is satisfiable, where $D'[x]$ is the translation of

D and P is predicate symbol not occurring in D'[x]. Moreover, D is universally satisfiable iff $((\forall x)P(x) \Leftrightarrow D'[x])$ And $(\forall x)P(x)$ is satisfiable.

We will present a sample T-box given by F. Baader.

female = (not male)
person = (some sex (or male female))
parent = (and person (some child person))
mother = (and parent (some sex female))
father = (and parent (not mother))
randparent = (and parent (some child parent))
parent–with–sons–only = (and parent (all child (some sex male)))

The following is a translation of this T-box into predicate logic:

$(\forall x)$ (female(x) $\Leftrightarrow$ $\neg$ male(x))

$(\forall x)$ (person(x) $\Leftrightarrow$ $(\exists y)$(sex(x,y) $\wedge$ (male(y) $\vee$ female(y))))

$(\forall x)$ (parent(x) $\Leftrightarrow$ person(x) $\wedge$ $(\exists y)$(child(x, y) $\wedge$ person(y)))

$(\forall x)$ (mother(x) $\Leftrightarrow$ parent(x) $\wedge$ $(\exists y)$(sex(x, y) $\wedge$ female(y)))

$(\forall x)$ (father(x) $\Leftrightarrow$ parent(x) $\wedge$ $\neg$ mother(x))

$(\forall x)$ (grandparent(x) $\Leftrightarrow$ parent(x) $\wedge$ $(\exists y)$(child(x, y) $\wedge$ parent(y)))

$(\forall x)$ (parent–with–sons–only(x) $\Leftrightarrow$ parent(x) $\wedge$

$\wedge$ $(\forall y)$(child(x, y) $\Rightarrow$ $(\exists z)$(sex(y, z) $\wedge$ male(z))))

8.1.2 Clause forms of ALC definitions

As the ALC-language allows arbitrary nesting of quantifiers, it is not immediately clear whether it corresponds to a (subset of) some well-known decision class. However, due to the fact that any subformula starting with a quantifier contains at most one free variable, the following operator *dissect* converts any predicate logic form of an ALC-definition to a new formula, belonging to a decidable class closely related with Gödel's Class (i. e. the class with prefix $\exists^*\forall\forall\exists^*$):

DEFINITION 8. 3.: Let F be the translation of some ALC-definition. Let G[x] be some subformula of F s.t. G starts with a quantifier which is in the scope of two other quantifiers. (Remember that the notation G[x] indicates that G contains a single free variable x). Then *dissect*(F) ≡ ($F^G_{P(x)}$ ∧ (∀x)(P(x) ⇔ G[x]), where $F^G_{P(x)}$ is the result of replacing in F the subformula G with the atom P(x) where P is a new predicate symbol. If there is no subformula of the indicated kind in F then let dissect(F) ≡ F.

We call a formula F* that results from iterated applications of dissect to a formula F simplified form of F iff dissect(F*) = F*.

Obviously, there is no unique simplified form in general. But the property we are interested in is captured by the following:

Proposition 8. 2.: A simplified form F* of a translation F of some ALC-definition is satisfiable iff F is satisfiable.

In the sections to come we will use clause sets corresponding to simplified forms of translations of ALC-definitions to demonstrate that questions concerning (universal) satisfiability and subsumption of concepts are decidable using resolution methods.

As an example, we will present a translation of the last formula from our first sample T-box (the other formulas remain unchanged):

(∀x)(parent–with–sons–only(x) ⇔ parent(x) ∧ (∀y)(child(x,y) ⇒ P_1(y)))
(∀y)(P_1(y) ⇔ (∃z)(sex(y,z) ∧ male(z)))

The following set *KL–ONE–example* is a clause form of the translated T-box above:

comment: definition of female:

{¬female(x), ¬male(x)}
{male(x), female(x)}

<u>comment:</u> definition of person:

{¬ person(x), sex(x,f(x))}

{¬ person(x), male(f(x)), female(f(x))}

{¬ sex(x,y), ¬ male(y), person(x)}

{¬ sex(x,y), ¬ female(y), person(x)}

<u>comment:</u> definition of parent:

{¬ parent(x), person(x)}

{¬ parent(x), child(x,g(x))}

{¬ parent(x), person(g(x))}

{¬ person(x), ¬ child(x,y), ¬ person(y), parent(x)}

<u>comment:</u> definition of mother:

{¬ mother(x), parent(x)}

{¬ mother(x), sex(x,h(x))}

{¬ mother(x), female(h(x))}

{¬ parent(x), ¬ sex(x,y), ¬ female(y), mother(x)}

<u>comment</u>: definition of father:

{¬ father(x), parent(x)}

{¬ father(x), ¬ mother(x)}

{¬ parent(x), mother(x), father(x)}

<u>comment</u>: definition of grandparent:

{¬ grandparent(x), parent(x)}

{¬ grandparent(x), child(x,k(x))}

{¬ granparent(x), parent(k(x))}

{¬ parent(x), ¬ child(x,y), ¬ parent(y), grandparent(x)}

<u>comment</u>: definition of parent–with–sons–only (abbreviated as pwso):

{¬ pwso(x), parent(x)}

{¬ pwso(x), ¬ child(x,y), p1(y)}

{¬ parent(x), child(x,l(x)), pwso(x)}

{¬ parent(x), ¬ p1(l(x)), pwso(x)}

{¬ p1(y), sex(y,r(y))}

{¬ p1(y), male(r(y))}

{¬ sex(y,z), ¬ male(z), p1(y)}

Notice that the descriptive power of a T-box above is more limited than one might expect. For example, given the following translation of an A-box

{person(John), child(John,Peter), sex(Peter,Male), male(Male)},

it is not possible to deduce pwso(John). In order to deduce pwso(John), we would need a fact $(\forall x)(\text{child}(\text{John},x) \Rightarrow x = \text{Peter})$. Unfortunately we do not have the equality predicate in the language!

Some experiments with the clause set above using an implementation of the k-refinement of chapter 4 are described in section 9.2.2 below.

8.1.3 Extensions of ALC

Consider the following natural extension (we call it *One-free*) of the class of formulas that are translations of ALC-definitions or descriptions.

DEFINITION 8.4 .: Any formula F of the predicate logic without the equality predicate and without function symbols belongs to the class *One-free* iff any subformula of F starting with a quantifier contains at most one free variable.

Recall that we do not consider the constant symbols to be functional - with other words, we allow formulas in *One-free* to contain constant symbols. The operator *dissect* obviously remains sound (w.r.t. to satisfiability) for class *One-free* and guarantees that no subformula of a simplified formula is in the scope of more than two quantifiers. Thus the resulting formulas are closely related with the formulas in the Gödel class (cf. chapter 5 or 6). We are not exactly concerned with the Gödel class, since the prenex form of a simplified formula may have an arbitrarily deep quantifier nesting.

However, it is easy to see that we may transform any simplified formula of class One-free to a set S of clauses s.t. for all $C \in S$: Either

(i) $|V(C)| \leq 2$ and C is function free, or

(ii) C contains only one variable x and at most unary function symbols all of which contain x as argument.

Observe that the resulting class of clause sets is a simple subset of class $\mathcal{S}^+$, described in chapter 5. Therefore we may use the resolution refinement R_m (see section 5.4.1 and 5.4.2) as a decision procedure for this class. Some simple additional observations thus imply that resolution methods may be employed to decide various important properties of ALC-definitions, like (universal) satisfiability and subsumption of concepts. If S is the clause form corresponding to the concept defintions of some A-box, $R_m^*(S)$ may be used as a basis to anwser efficiently whole series of questions about this A-box.

To mention another resolution variant that is investigated in this monography we remark that also the termination of the k-refinement of chapter 4 (or $>_{sv}$-refinement of chapter 5) on this class is easy to show: One has to modify the termination proof for Gödel's Class. An important point to notice is that any clause derived from a clause obeying (i) above and a clause obeying (ii) is splittable into clauses obeying (ii). Analogously to the Gödel class, there is no need for actual splitting during proof search - using Robinson's resolution or corresponding restriction on factorization is enough.

As for completeness, the same remarks hold as for Gödel's class: to preserve lifting, subsumption must be restricted in the following way: a literal P(x, y) must not be allowed to subsume a literal P(x, x){t/x} for any t. Tautology-elimination must be restricted correspondingly:

{P(x,x), ¬P(x,x)} {t/x} should not be eliminated for any t, while eliminating {P(x,y), ¬P(x,y)} is OK.

8.1.4 Functional relations

Some extended versions of the ALC language (the one presented in [HN 90], for example) also allow to declare certain relations to be functional (e.g., the relation "sex" in the example above would naturally be defined as a functional relation). The predicate logic equivalent of declaring a relation (say, R) to be functional on the first argument would be a following axiom:

$$(\forall xyz)\ ((R(x,y) \wedge R(x,z)) \Rightarrow (y = z))$$

The axiom uses the equality predicate and thus we would have to handle the defining properties of equality e.g. by explicit axiomatization of equality, that is introduction of axioms for reflexivity, commutativity, transitivity and axioms for substitution into atoms. Including the full axiomatization for equality would mean that the resulting formulas are not any more members of known decidable classes of predicate logic, thus making the translation useless for our purposes.

In principle functionality of a binary relation can be naturally expressed using one-place function symbols instead of two-place predicate symbols combined with the functionality axiom. This means that instead of using a quantifier to introduce a new functionally bound variable, we can use a term f(x), where f is a function symbol representing the functional relation and x depends on the context.

So, we could consider extending class *One-free* by allowing one-place function symbols. Unfortunately, this extension turns out to be undecidable, as it contains the $\forall\exists\forall$-Class.

We could consider ways of imposing additional restrictions to the extended *One-free* class. One rather strong restriction, which we will shortly investigate in the following, is obtained by disallowing any non-functional relations. That is, we require that all binary relations are functional - then they can be represented by one-place function symbols and we can restrict the class to contain only monadic predicate symbols. The resulting class, of course, is not any longer an extension but rather a variant of class *One-free*.

It is natural to extend this new class in a way s.t. it corresponds to class E (see chapters 4 and 5).

<u>DEFINITION</u> 8.5.: Any formula F of the predicate logic without the equality predicate, but possibly containing function symbols belongs to the Class *One-free-E* (OFE, for short) iff any subformula of F starting with a quantifier contains at most one free variable and all atoms contain at most one variable.

Obviously the simplified form (using the operator simple) of any formula in class OFE belongs to the Class E, the decidability of which has been shown in various ways in chapters 4 and 5.

Class OFE can be used as a media to express questions about both, the T-box and the A-box. As for the A-box, we also allow ground equality units such that the set of all equality units can be oriented to a complete (confluent and terminating) set of term rewriting rules (cf. [KBR86]. Thus all the facts of the form R(a,b) for any functional R can be translated to equality units f(a) = b.

In order to decide any formula F in the OFE Class containing also ground equality units, do the following:

- Convert F to a simplified form F^* (using the operater *dissect* presented in the last section).
- Convert F^* to a clause set S (equivalent to F^* w.r.t. satisfiability).
- Use the Knuth-Bendix algorithm to check the completeness of the set of equality units; if needed, try to complete the set. Notice that on the ground set of equality units the Knuth-Bendix completion algorithm will always terminate.
- Use narrowing to remove the rewrite rules resulting from the previous step - this gives a fully narrowed clause set S' which is in class E. (cf. [Sla74] and chapter 8 to see the definition and examples of narrowings.)
- Use techniques from chapter 4 or 5 to decide the clause set S'.

As an example we take the previous T-box, assume the relation "sex" to be declared functional and translate the selected part to OFE. Unfortunately, the OFE Class is too restrictive to translate the whole of the previous T-box in a "coherent" way. The reason here is that the relation child cannot be considered to be functional on the first argument. We will use the notion of "first child" instead: ***first–child*** is assumed to be functional. Whenever some person x does not have children, we can define first–child(x) = ⊥, for some constant ⊥ for which we know that ¬person(⊥). In the following T-box we have replaced some equivalences by implications - the sole reason being that we felt the implications to correspond better to our intuitive understanding of the presented terminology.

$\neg$ person($\bot$)

female(Female)

male(Male)

$(\forall x)$ (female(x) $\Rightarrow$ $\neg$ male(x))

$(\forall x)$ (person(x) $\Rightarrow$ male(sex(x)) $\vee$ female(sex(x)))

$(\forall x)$ (parent(x) $\Leftrightarrow$ person(x) $\wedge$ person(first–child(x)))

$(\forall x)$ (mother(x) $\Leftrightarrow$ parent(x) $\wedge$ female(sex(x)))

$(\forall x)$ (father(x) $\Leftrightarrow$ parent(x) $\wedge$ $\neg$ mother(x))

As for the concepts *grandparent* and *parent–with–sons–only*, we could assume that the number of children a person can have is bounded by some N and use the following clumsy translation:

$(\forall x)$ (grandparent(x) $\Leftrightarrow$ parent(x) $\wedge$

(parent(first–child(x)) $\vee$ ... $\vee$ parent(N^{th}–child(x))))

$(\forall x)$ (man(x) $\Leftrightarrow$ (person(x) $\wedge$ male(sex(x))))

$(\forall x)$ (parent–with–sons–only(x) $\Leftrightarrow$ parent(x) $\wedge$

$\bigwedge_{i=1}^{N}$ person(i^{th}child) $\Rightarrow$ male(sex(i^{th} child)))

Now we could, for example, add a sample A-box to the given T-box (John, Peter and Male are constant symbols):

first–child(John) = Peter

second–child(John) = $\bot$, ... , N^{th}–child(John) = $\bot$

sex(first–child(John)) = Male

grandparent(John)

Completion will add a new equality unit sex(Peter) = Male to the ones above. Now it is easy to check out that from the full narrowing of the clause form of the last T-box above it is possible to derive e.g. father(Peter) without using the equality units any more. The following clauses from the fully narrowed set are used for this derivation:

{$\neg$ person($\bot$)}

{male(Male)}

{$\neg$ female(x), $\neg$ male(x)}

{$\neg$ mother(Peter), female(Male)}

{$\neg$ parent(x), mother(x), father(x)}

{¬grandparent(John), parent(Peter), ... , parent(⊥)}
{¬parent(x), person(x) }
{grandparent(John)}

8.2 EXPERIMENTS WITH THE IMPLEMENTATION

In order to experiment with the decision refinements of resolution, we have written a resolution theorem prover containing various well-known and less well-known search strategies and refinements. The prover (see [Tam 90], [Tam91]) is implemented on IBM-PC in muLISP; unification and subsumption are coded directly in assembly. The prover is suitable both for fully automatic and interactive use.

The prover incorporates a special part for building finite models. This part does not implement the complete method described in chapter 8, but a "minimal consistent criteria" method with no backtracing. We do not have a proof that such a simple method always terminates on Class AM; neither have we any counterexamples. The minimal-criteria method has performed quite well in the experiments. Although we know that it does not terminate on all formulas from Gödel's Class, it has been able to build finite models for several satisfiable formulas in Gödel's Class presented in [Chu56].

The prover has been used as an aid for solving an open problem: condensed detachment completeness of relevance logic (see [MT91]). However, decision strategies were not used for this application and thus we won't give any details here.

All the following experiments have been performed on the 640-Kbyte machine with the 80286 processor running at 16 MHZ - a rather small and slow machine by current standards.

8.2.1 Hyperresolution-based methods

In chapter 3 semantic clash refinements where used to decide the classes PVD and OCC1N. The essential feature of the decision method was the algorithmical selection of an appropriate setting according to the syntactical structure of the set of clauses. We will show that the decision method for PVD is also an efficient theorem prover.

Consider first the set of clauses

$$H_1 = \{\{\neg P(x,y,z),\ P(z,y,x)\}$$
$$\{\neg P(x,y,z),\ P(y,x,z)\}$$
$$\{\neg P(x,y,z),\ P(x,y,g(z))\}$$
$$\{\neg P(x,y,z),\ P(x,y,f(z))\}$$
$$\{P(a,b,c)\}$$
$$\{\neg P(f(g(a)),\ f(g(b)),\ f(g(c)))\}\}$$

It is easy to see that H_1 is in PVD via the positive setting M_P. Using negative hyperresolution (i.e. computing $R^{*}_{M_P}(H_1)$) the empty clause was derived in 14 seconds on level 12, 162 derived clauses were retained. By "retained" we mean "not immediately eliminated as tautologies or subsumed clauses". Thus, clauses eliminated by back subsumption are considered to be "retained".

Positive hyperresolution ran out of space at level 7, after retaining 1433 clauses in 450 seconds. The reason for the extremely different behaviour of positive and negative hyperresolution is the following: While negative hyperresolution reduces the depth of terms, positive hyperresolution does the contrary. The resolution based decision procedure for PVD, defined in chapter 3, automatically selects the positive setting and gets a fast refutation of H_1.

The following set H_2 is slightly more complex than H_1, but expresses a similiar mathematical problem.

$$H_2 = \{\{P(x,y,z,u),\ \neg P(u,x,y,z)\}$$
$$\{P(x,y,z,u),\ \neg P(u,z,y,x)\}$$
$$\{P(x,y,z,u),\ \neg P(f(x),y,z,u)\}$$
$$\{P(x,y,z,u),\ \neg P(x,y,z,g(u))\}$$
$$\{\neg P(a,b,c,d)\}$$
$$\{P(f(g(d)),f(g(c)),f(g(b)),f(g(a)))\}\}$$

H_2 also belongs to Class PVD, but via negative setting. We observe an effect similiar to that for H_1: By positive hyperresolution the refutation was found in 120 seconds on the 13th level. 645 clauses were retained. Negative hyperresolution ran out of space at level 6, after retaining 1339 clauses in 411 seconds.

While H_1 is in KII, H_2 is in KI (see chapter 3 and [Lei90]); but both, H_1 and H_2, are in PVD (by KI $\cup$ KII $\subset$ PVD). While positive hyperresolution decides KI, negative decides KII, but no fixed setting refinement decides both of them. Moreover, by the experiments discussed before, we see that neither positive nor negative hyperresolution is suited for refuting both, H_1 and H_2. Thus the clause sets H_1,

H_2 illustrate the value of using resolution decision procedures as "ordinary" theorem provers: besides extension of the knowledge about the decision problem, finding larger decidable classes leads to better theorem proving programs. That a "decider" yields a good theorem prover is by no means surprising, because many resolution based decision procedures (as the one described for PVD) do not increase the term depth of clauses and thus keep complexity low.

In general it will pay out to build an expert system, which - given a set of clauses S - tries to find a (predefined) decision class Γ s.t. $S \in \Gamma$; afterwards the decision method is used to refute S. By the undecidability of predicate logic such a method cannot always succeed, but may be of practical value nevertheless.

Finally we define a set of clauses H_3 which is not in PVD (although again "similiar" to H_1 and H_2).

$$
\begin{aligned}
H_3 = \{&\{\neg P(x,y,z,), \ \neg P(y,u,z), \ P(x,u,z)\} \\
&\{\neg P(x,y,z), \ P(z,y,x)\} \\
&\{\neg P(x,y,z), \ P(y,x,z)\} \\
&\{\neg P(x,y,z), \ P(x,y,f(z))\} \\
&\{\neg P(x,y,z), \ P(g(x),y,z)\} \\
&\{P(a,f(b),c)\} \\
&\{P(f(b),d,c)\} \\
&\{\neg \ P(g(f(a)), g(f(d)), g(f(c)))\}\}
\end{aligned}
$$

Without the first clause $C = \{\neg P(x, y, z), \neg P(y, u, z), P(x, u, z)\}$ H_3 would be in PVD via positive setting. But for M_p we get $V(C_{neg}) - V(C_{pos}) \neq \emptyset$, which violates PVD2. On the other hand, taking the negative setting M_n and the clause $D = \{\neg P(x,y,z), P(x,y,f(z))\}$ we get $\tau_{MAX}(x, D_{neg}) > \tau_{MAX}(x, D_{pos})$, which again violates PVD2. Thus no set of clauses containing C as well as D can be in PVD. It is significant that both, positive hyperresolution and negative hyperresolution ran out of space in the attempt to refute H_3. Moreover, it is easy to verify that also A-ordering refinements fail to refute H_3 within reasonable time and space bounds. In fact H_3 is not contained in any decidable class investigated in this monography. However, it is not hard to find a hyperresolution refutation of depth 14 by hand (transitivity and permutations are easily handled by human problem solvers, but not by standard resolution theorem provers).

8.2.2 Ordering refinements

[Chu 56] presents a large set of formulas in several decidable classes, meant as exercises for the reader. Some of these formulas are satisfiable, some are not. Some of the formulas are very easy to decide, some are quite hard.

The following table presents the results of using the k-refinement for deciding all the formulas from [Chu 56]. Each row of the table contains the number of the formula (in the set of exercises to section 46 of [Chu 56]), its prefix, time of deciding a formula, number of derived clauses, number of retained clauses, result found (negation of the formula satisfiable-/unsatisfiable), time of constructing a finite domain, size of the domain found. All times are given in seconds (shorter times are somewhat inaccurate). '***' means that the finite model was not found (due to memory limitations or inherent incompleteness of the used model-building method).

Ex2 No1	EAE	0.11	3	2	unsat.	-	-
Ex2 No2	AE	0.06	1	0	sat.	0.05	2
Ex2 No3	EEEA	0.02	0	0	sat.	0.06	1
Ex2 No4	EEA	0.11	1	1	unsat.	-	-
Ex2 No5	AAEE	0.38	13	9	unsat.	-	-
Ex3 No1	EAEE	0.05	0	0	sat.	0.17	2

Ex3 No2	EEEAA	0.11	3	2	unsat.	-	-
Ex4 No1	EA	0.05	2	1	sat.	0.04	1
Ex4 No2	EA	0.17	3	2	unsat.	-	-
Ex9 No1	AE	0.17	8	2	sat.	0.55	2
Ex9 No2	AE	0.22	5	5	unsat.	-	-
Ex12 No1	AEE	0.33	6	6	unsat.	-	-
Ex12 No2	EAE	0.33	16	5	unsat.	-	-
Ex12 No3	AEE	1.93	112	40	unsat.	-	-
Ex14 No1	AAE	0.06	0	0	sat.	4.5	7
Ex14 No2	AAE	2.69	196	20	sat.	***	***
Ex14 No3	AAE	0.22	3	3	unsat.	-	-
Ex14 No4	AAE	2.75	214	30	unsat.	-	-
Ex14 No5	AAE	0.28	5	5	unsat.	-	-
Ex14 No6	AAE	1.43	93	21	unsat.	-	-
Ex14 No7	AAE	0.55	19	5	sat.	***	***
Ex15 No1	EAAE	0.11	0	0	unsat.	-	-
Ex15 No2	EEAE	0.11	0	0	sat.	0.05	2
Ex15 No3	AEA	0.06	0	0	unsat.	-	-
Ex15 No4	AEA	0.06	0	0	unsat.	-	-
Ex15 No5	AEAA	0.05	0	0	unsat.	-	-
Ex15 No6	AEAA	0.05	0	0	unsat.	-	-
Ex15 No7	EAEA	0.06	0	0	sat.	0.06	2
Ex16 No2	EAAE	0.06	0	0	unsat.	-	-
Ex16 No3	EAAE	0.11	1	1	sat.	0.17	2
Ex16 No4	EEAAE	0.72	9	5	unsat.	-	-
Ex17 No2	EEAAEE	0.11	1	1	unsat.	-	-
Ex17 No3	EEAAE	1.32	42	34	unsat.	-	-
Ex17 No4	AEAE	30.5	3182	0	sat.	***	***
Ex17 No5	AEAE	7.58	588	22	unsat.	-	-
Ex18 No2	EAAE	1.43	98	10	unsat.	-	-
Ex18 No3	EEAAE	2.04	140	33	sat.	8.79	4
Ex18 No4	EEAAE	0.66	29	8	unsat.	-	-
Ex18 No5	EAAAE	15.16	654	109	unsat.	-	-
Ex20 No1	EEAAE	0.66	28	11	unsat.	-	-

An important point noticed during experimentation is that the decision refinement performed very well (when compared to any other known resolution strategy) for the unsatisfiable formulas in [Chu 56]. Consider the hardest unsatisfiable formula in [Chu 56]: Ex18 No5. The following is the clause form of Ex18 No5:

$\{\neg P(x,y,z),\ \neg P(a,a,f(x,y,z)),\ P(y,z,x),\ P(z,x,y)\}$

$\{\neg P(z,x,y),\ P(x,y,z),\ \neg P(y,x,f(x,y,z))\}$

$\{\neg P(z,x,y),\ P(y,z,x),\ \neg P(y,x,f(x,y,z))\}$

$\{P(z,x,y),\ P(y,x,f(x,y,z))\}$

$\{\neg P(x,y,z),\ \neg P(y,z,x),\ P(y,x,f(x,y,z))\}$

$\{\neg P(y,z,x),\ P(x,y,z),\ \neg P(x,f(x,y,z),y)\}$

$\{\neg P(y,z,x),\ P(z,x,y),\ \neg P(x,f(x,y,z),y)\}$

$\{P(y,z,x),\ P(x,f(x,y,z),y)\}$

$\{\neg P(x,y,z),\ \neg P(z,x,y),\ P(x,f(x,y,z),y)\}$

$\{P(z,x,y),\ P(x,y,z),\ \neg P(f(x,y,z),y,x)\}$

$\{P(y,z,x),\ P(x,y,z),\ \neg P(f(x,y,z),y,x)\}$

$\{\neg P(z,x,y),\ \neg P(y,z,x),\ P(f(x,y,z),y,x)\}$

$\{\neg P(x,y,z),\ P(f(x,y,z),y,x)\}$

$\{P(x,y,z),\ P(f(x,y,z),f(x,y,z),f(x,y,z))\}$

$\{P(y,z,x),\ P(f(x,y,z),f(x,y,z),f(x,y,z))\}$

$\{P(z,x,y),\ P(f(x,y,z),f(x,y,z),f(x,y,z))\}$

$\{\neg P(x,y,z),\ \neg P(y,z,x),\ \neg P(z,x,y),\ \neg P(f(x,y,z),f(x,y,z),f(x,y,z))\}$

The derivation of the empty clause found by the prover was 26 levels deep. With no other well-known strategy of resolution was the prover able to show unsatifiability of this clause set. We experimented also with W. McCune's prover OTTER, and this prover was also unable to find the refutation. Of course, all the experiments were performed on a rather small machine. It would be interesting to see how powerful a machine is needed to find the refutation of the clause set above using some proof strategy sufficiently different from the ordering refinement we used.

8.2.2 Experiments with the translations of KL-ONE formulas

Recall the example clause set *KL–ONE–example* from the above section describing the KL-ONE type languages. Our implementation of the k-refinement was able to show the satisfiability of this set in 10 seconds. The whole derivation was 8 levels deep, 320 clauses were derived and 70 of those were retained. The program was also successful to find out that this clause set has a finite model with an 1-element domain.

For this particular example the satisfiability follows from the propositional structure already and positive hyperresolution is unable to derive a single new clause from the whole set. However, the k-refinement did not use this fact. Also, it can be considered rather important to find the "completed set" and the whole KL-ONE-subsumption structure for the formula. The last two tasks were performed during those 10 seconds of work of the prover.

REFERENCES

[BBHNS90] F. Baader, H. J. Burckert, B. Hollunder, W. Nutt, J. H. Siekmann: Concept Logics. Research report RR-90-10, Deutsches Forschungszentrum für Künstliche Intelligenz GmbH, Kaiserslautern, Germany, 1990.

[BBMR89] A. Borgida, R. J. Brachmann, D. L. McGuinness, L. A. Resnick: CLASSIC: A Structural Data Model for Objects. Proceedings of the 4th National Conference of the AAAI, Austin, Texas, 1984, pp. 34-37.

[BPGL85] R. J. Brachmann, G. Pigman, H. J. Levesque: An Essential Hybrid Reasoning System: Knowledge and Symbol Level Accounts in KRYPTON. Proceedings of the 9th IJCAI, Los Angeles, California, 1985, pp. 532-539.

[BS28] P. Bernays, M. Schönfinkel: Zum Entscheidungsproblem der mathematischen Logik. Math. Ann. 99 (1928), pp. 342-372.

[BS85] R. J. Brachmann, J. G. Schmolze: An Overview of the KL-ONE Knowledge Representation System. Cognitive Science 9(2), 1985, pp. 171-216.

[Bar 81] H. P. Barendregt: The Lambda Calculus. (North Holland, Amsterdam, 1981).

[Boy 71] R. S. Boyer: Locking: A Restriction of Resolution. The University of Texas at Austin, Ph.D. dissertation, 1971.

[CGT 90] S. Ceri, G. Gottlob, L. Tanca: Logic Programming and Databases. Springer, Berlin Heidelberg New York 1990.

[CL 73] C. L. Chang, R.C.T. Lee: Symbolic Logic and Automated Theorem Proving. Academic press, 1973.

[Chu 36] A. Church: A Note on the Entscheidungsproblem. Journal of Symbolic Logic 1 (1936) pp. 40 - 44.

[Chu 56] A. Church: Introduction to Mathematical Logic I. (Princeton University Press, New Jersey, 1956).

[DL83] L. Denenberg, R. Lewis: Logical Syntax and Computational Complexity. Proc. Logic Colleg '83. LNCS 104, pp. 101-115.

[DG79] B. Dreben, W.D. Goldfarb: The Decision Problem: Solvable Classes of Quantificational Formulas. (Addison-Wesley, Massachusetts 1979).

[DP60] M. Davis, H. Putnam: A Computing Procedure for Quantification Theory. Journal of the ACM 7 No. 3, 1960, pp. 201 - 215.

[Fer90] C. Fermüller: Deciding some Horn Clause Sets by Resolution. Yearbook 1989 of the Kurt Gödel Society, Vienna 1990, pp. 60 - 73.

[Fer91] C. Fermüller: Deciding Classes of Clause Sets by Resolution. PhD - Thesis, Technical University of Vienna 1991.

[Fer91a] C. Fermüller: A Resolution Variant Deciding some Classes of Clause Sets. CSL'90, LNCS 533, 1991, pp. 128 - 144.

[Gur73] Y. Gurevich: Formuly s odnim $\forall$ (formulas with one $\forall$. In Izbrannye voprosy algebry i logiki (Selected Questions in Algebra and Logics; in memory of A. Mal'cev). Nauka, Nowosibirsk, 1973, pp. 97-110.

[Hil01] D. Hilbert: Probleme der Mathematik. Archiv der Mathematik und Physik (3)/1, 1901, pp. 44 - 63, 213 - 237.

[HN90] B. Hollunder, W. Nutt: Subsumption Algorithms for Concept Languages. Research report RR-90-04, Deutsches Forschungszentrum für Künstliche Intelligenz GmbH, Kaiserslautern, Germany, 1990.

[Joy73] W. H. Joyner: Automated Theorem Proving and the Decision Problem. Ph.D.Thesis, Harvard University, May 1973.

[Joy76] W. H. Joyner: Resolution Strategies as Decision Procedures. J. ACM 23, 1 (July 1976), pp. 398-417.

[KBR86] T. S. Kaczmarek, R. Bates, G. Robins: Recent Developments in NIKL. Proceedings of the 5th National Conference of the AAAI, Philadelphia, PA, 1986, pp. 578-587.

[KH69] R. Kowalski, P.J. Hayes: Semantic Trees in Automated Theorem Proving; in Machine Intelligence 4, B. Meltzer and D. Michie, Eds., Edinburgh U. Press, Edinburgh, 1969, pp. 87-101.

[LG 90] A. Leitsch, G. Gottlob: Deciding Horn Clause Implication Problems by Ordered Semantic Resolution. Computational Intelligence II, North Holland 1990, pp. 19 - 26.

[Leib] G. W. Leibniz: Calculus Ratiocinator; in: sämtliche schriften und Briefe; herausgegeben von der Preußischen Akademie der Wissenschaften Darmstadt, Reichel, 1923.

[Lei 88] A. Leitsch: Implication Algorithms for Classes of Horn Clauses. Statistik, Informatik + Oekonomie, Springer Verl. Berlin, Heidelberg, 1988, pp. 172-189.

[Lei 89] A. Leitsch: On Different Concepts of Resolution. Zeitschr. für Math. Logik und Grundlagen der Mathematik 35, 1989, pp. 71 - 77.

[Lei 90] A. Leitsch: Deciding Horn Classes by Hyperresolution. CSL' 89, Lecture Notes in Computer Science 440, pp. 225 - 241.

[Lew79] H. R. Lewis: Unsolvable Classes of Quantificational Formulas. Addison Wesley, Massachusetts 1979.

[Lew 80] H. R. Lewis: Complexity Results for Classes of Quantificational Formulas. J. Computer and System Sciences 21 (3) (1980) pp. 317 - 353.

[Lov 78] D. Loveland: Automated Theorem Proving - A Logical Basis. North Holland Publ. Comp. 1978.

[Löw15] L. Löwenheim: Über Möglichkeiten im Relativkalkül. Math. Annalen 68, pp. 169-207.

[MB 87] R. MacGregor, R. Bates: The Loom Knowledge Representation Language. Technical Report ISI/RS-87-188, University of Southern Carolina, Information Science Institute, Marina del Rey, California, 1987.

[MP92] J. Marcinkowski, L. Pacholski: Undecidability of the Horn-Clause Implication Problem. Rapport de Recherce 1992 - 5, Groupe de recherche algorithmique & logique, University of Caen, France 1992.

[MT 91] G. Mints, T. Tammet: Condensed Detachment is Complete for Relevance Logic: A Computer Aided Proof. Journal of Automated Reasoning 7, 1991, pp. 587 - 596.

[Mas 64] S.Y. Maslov: An Inverse Method of Establishing Deducibilities in the Classical Predicate Calculus. Dokl. Akad.Nauk SSSR 159, pp. 1420 - 1424.

[Mas 68] S.Y. Maslov: The Inverse Method for Establishing Deducibility for Logical Calculi. Trudy Mat. Inst. Steklov 98 (1968) 26-87= Proc. Steklov. Inst. Math. 98 (1968) 25-96, MR 40 #5416; 43 #4620.

[Mas 71] S.Y. Maslov: Proof-Search Strategies for Methods of the Resolution Type. Machine Intelligence 6 (American Elsevier, 1971, pp. 77 - 90.

[Nol 80] H. Noll: A Note on Resolution: How to Get Rid of Factoring Without Loosing Completeness. 5th Conference on Automated Deduction, Lecture Notes in Computer Science 87, pp. 250 - 263.

[NvL 88] B. Nebel, K. von Luck: Hybrid Reasoning in BACK. In Z.W. Ras, L. Saitta (editors): Methodologies for Intelligent Systems, North Holland, Amsterdam, Netherlands, 1988, pp. 260 - 269.

[Pat 84] P. Patel-Schneider: Small can be Beautiful in Knowledge Representation. Proceedings of the IEEE Workshop on Principles of Knowledge-based Systems, Denver, Colorado, 1984, pp. 11 - 16.

[Rob 63] J.A. Robinson: Theorem Proving on the Computer. J.ACM. 10 No 2, 1963, pp. 163 - 174.

[Rob 65] J. A. Robinson: A Machione Oriented Logic Based on the Resolution Principle. Journal of the ACM 12, 1965, pp. 23 - 41.

[Rob65a] J. A. Robinson: Automated Deduction with Hyperresolution. Intern. Journal of Computer Mathematics 1, 1965, pp. 227 - 334.

[Rob68] J. A. Robinson: The Generalized Resolution Principle. Machine Intelligence 3, Edinburgh, 1968, pp. 77 - 93.

[Rud 91] V. Rudenko: A new Result on the Horn Implication Problem. Yearbook 1992 of the Kurt - Gödel - Society. Vienna, 1992.

[SS 88] M. Schmidt-Schauss, A. Smolka, G. Attributive Concept Descriptions with Unions and Complements. SEKI report SR-88-21, FB Informatik, Universität Kaiserslautern, Germany, 1988.

[Sch 88] M. Schmidt-Schauss: Implication of Clauses is Undecidable. Theoretical Computer Science 59, 1988, pp. 287 - 296.

[Sla 67] J. R. Slagle: Automatic Theorem Proving with Renamable and Semantic Resolution. Journal of the ACM 14 No 4, 1967, pp. 687 - 697.

[Sla 74] J. R. Slagle: Automated Theorem-Proving for Theories with Simplifiers, Commutativity and Associativity. J. ACM 21 (4), 1974, pp. 622 - 642.

[Ste 71] S. Stenlund: Combinators, λ-Terms and Proof Theory. Reidel Publ. Comp. 1971.

[Tam 90] T. Tammet: The Resolution Program, Able to Decide some Solvable Classes. COLOG-88, Springer LNCS 417, 1990, pp. 300 - 312.

[Tam 91] T. Tammet: Using Resolution for Deciding Solvable Classes and Building Finite Models. Baltic Computer Science, LNCS 502, 1991, pp. 33 - 64.

[Vil 85] M. B. Vilain: The Restricted Language Architecture of a Hybrid Representation System. In R.J. Bachmann, H.J. Levesque, R. Reiter (editors): Proceedings of the 9th IJCAI, Los Angeles, California, 1985, pp. 547 - 551.

[WOLB84] L. Wos, R. Overbeek, E. Lusk, J. Boyle: Automated Reasoning: Introduction and Applications. (Prentice - Hall, New Jersey, 1984).

[ZS 74] N. K. Zamov, V. I. Sharonov: Decision Tactics of Proof Search for the Resolution Method. 1974. In Russian.

[Zam 72] N. K. Zamov: On a Bound for the Complexity of Terms in the Resolution Method. Trudy Mat. Inst. Steklov 128, 1972, pp. 5 - 13.

[Zam 89] N. K. Zamov: Local Search Methods in Automated Theorem Proving. Dissertation for D.Sc., Kazan 1989. In Russian.

[Zam89a] N. K. Zamov: Maslov's Inverse Method and Decidable Classes. Annals of Pure and Applied Logic 42, 1989, pp. 165 - 194.

INDEX

Printing: Weihert-Druck GmbH, Darmstadt
Binding: Buchbinderei Schäffer, Grünstadt

Lecture Notes in Artificial Intelligence (LNAI)

Vol. 517: K. Nökel, Temporally Distributed Symptoms in Technical Diagnosis. IX, 164 pages. 1991.

Vol. 518: J. G. Williams, Instantiation Theory. VIII, 133 pages. 1991.

Vol. 522: J. Hertzberg (Ed.), European Workshop on Planning. Proceedings, 1991. VII, 121 pages. 1991.

Vol. 535: P. Jorrand, J. Kelemen (Eds.), Fundamentals of Artificial Intelligence Research. Proceedings, 1991. VIII, 255 pages. 1991.

Vol. 541: P. Barahona, L. Moniz Pereira, A. Porto (Eds.), EPIA '91. Proceedings, 1991. VIII, 292 pages. 1991.

Vol. 542: Z. W. Ras, M. Zemankova (Eds.), Methodologies for Intelligent Systems. Proceedings, 1991. X, 644 pages. 1991.

Vol. 543: J. Dix, K. P. Jantke, P. H. Schmitt (Eds.), Nonmonotonic and Inductive Logic. Proceedings, 1990. X, 243 pages. 1991.

Vol. 546: O. Herzog, C.-R. Rollinger (Eds.), Text Understanding in LILOG. XI, 738 pages. 1991.

Vol. 549: E. Ardizzone, S. Gaglio, F. Sorbello (Eds.), Trends in Artificial Intelligence. Proceedings, 1991. XIV, 479 pages. 1991.

Vol. 565: J. D. Becker, I. Eisele, F. W. Mündemann (Eds.), Parallelism, Learning, Evolution. Proceedings, 1989. VIII, 525 pages. 1991.

Vol. 567: H. Boley, M. M. Richter (Eds.), Processing Declarative Kowledge. Proceedings, 1991. XII, 427 pages. 1991.

Vol. 568: H.-J. Bürckert, A Resolution Principle for a Logic with Restricted Quantifiers. X, 116 pages. 1991.

Vol. 587: R. Dale, E. Hovy, D. Rösner, O. Stock (Eds.), Aspects of Automated Natural Language Generation. Proceedings, 1992. VIII, 311 pages. 1992.

Vol. 590: B. Fronhöfer, G. Wrightson (Eds.), Parallelization in Inference Systems. Proceedings, 1990. VIII, 372 pages. 1992.

Vol. 592: A. Voronkov (Ed.), Logic Programming. Proceedings, 1991. IX, 514 pages. 1992.

Vol. 596: L.-H. Eriksson, L. Hallnäs, P. Schroeder-Heister (Eds.), Extensions of Logic Programming. Proceedings, 1991. VII, 369 pages. 1992.

Vol. 597: H. W. Guesgen, J. Hertzberg, A Perspective of Constraint-Based Reasoning. VIII, 123 pages. 1992.

Vol. 599: Th. Wetter, K.-D. Althoff, J. Boose, B. R. Gaines, M. Linster, F. Schmalhofer (Eds.), Current Developments in Knowledge Acquisition - EKAW '92. Proceedings. XIII, 444 pages. 1992.

Vol. 604: F. Belli, F. J. Radermacher (Eds.), Industrial and Engineering Applications of Artificial Intelligence and Expert Systems. Proceedings, 1992. XV, 702 pages. 1992.

Vol. 607: D. Kapur (Ed.), Automated Deduction – CADE-11. Proceedings, 1992. XV, 793 pages. 1992.

Vol. 610: F. von Martial, Coordinating Plans of Autonomous Agents. XII, 246 pages. 1992.

Vol. 611: M. P. Papazoglou, J. Zeleznikow (Eds.), The Next Generation of Information Systems: From Data to Knowledge. VIII, 310 pages. 1992.

Vol. 617: V. Mařík, O. Štěpánková, R. Trappl (Eds.), Advanced Topics in Artificial Intelligence. Proceedings, 1992. IX, 484 pages. 1992.

Vol. 619: D. Pearce, H. Wansing (Eds.), Nonclassical Logics and Information Processing. Proceedings, 1990. VII, 171 pages. 1992.

Vol. 622: F. Schmalhofer, G. Strube, Th. Wetter (Eds.), Contemporary Knowledge Engineering and Cognition. Proceedings, 1991. XII, 258 pages. 1992.

Vol. 624: A. Voronkov (Ed.), Logic Programming and Automated Reasoning. Proceedings, 1992. XIV, 509 pages. 1992.

Vol. 627: J. Pustejovsky, S. Bergler (Eds.), Lexical Semantics and Knowledge Representation. Proceedings, 1991. XII, 381 pages. 1992.

Vol. 633: D. Pearce, G. Wagner (Eds.), Logics in AI. Proceedings. VIII, 410 pages. 1992.

Vol. 636: G. Comyn, N. E. Fuchs, M. J. Ratcliffe (Eds.), Logic Programming in Action. Proceedings, 1992. X, 324 pages. 1992.

Vol. 638: A. F. Rocha, Neural Nets. A Theory for Brains and Machines. XV, 393 pages. 1992.

Vol. 642: K. P. Jantke (Ed.), Analogical and Inductive Inference. Proceedings, 1992. VIII, 319 pages. 1992.

Vol. 659: G. Brewka, K. P. Jantke, P. H. Schmitt (Eds.), Nonmonotonic and Inductive Logic. Proceedings, 1991. VIII, 332 pages. 1993.

Vol. 660: E. Lamma, P. Mello (Eds.), Extensions of Logic Programming. Proceedings, 1992. VIII, 417 pages. 1993.

Vol. 667: P. B. Brazdil (Ed.), Machine Learning: ECML – 93. Proceedings, 1993. XII, 471 pages. 1993.

Vol. 671: H. J. Ohlbach (Ed.), GWAI-92: Advances in Artificial Intelligence. Proceedings, 1992. XI, 397 pages. 1993.

Vol. 679: C. Fermüller, A. Leitsch, T. Tammet, N. Zamov, Resolution Methods for the Decision Problem. VIII, 205 pages. 1993.

Vol. 689: J. Komorowski, Z. W. Raś (Eds.), Methodologies for Intelligent Systems. Proceedings, 1993. XI, 653 pages. 1993.

Vol. 695: E. P. Klement, W. Slany (Eds.), Fuzzy Logic in Artificial Intelligence. Proceedings, 1993. VIII, 192 pages. 1993.

Lecture Notes in Computer Science

Vol. 658: R. A. Rueppel (Ed.), Advances in Cryptology – EUROCRYPT '92. Proceedings, 1992. X, 493 pages. 1993.

Vol. 659: G. Brewka, K. P. Jantke, P. H. Schmitt (Eds.), Nonmonotonic and Inductive Logic. Proceedings, 1991. VIII, 332 pages. 1993. (Subseries LNAI).

Vol. 660: E. Lamma, P. Mello (Eds.), Extensions of Logic Programming. Proceedings, 1992. VIII, 417 pages. 1993. (Subseries LNAI).

Vol. 661: S. J. Hanson, W. Remmele, R. L. Rivest (Eds.), Machine Learning: From Theory to Applications. VIII, 271 pages. 1993.

Vol. 662: M. Nitzberg, D. Mumford, T. Shiota, Filtering, Segmentation and Depth. VIII, 143 pages. 1993.

Vol. 663: G. v. Bochmann, D. K. Probst (Eds.), Computer Aided Verification. Proceedings, 1992. IX, 422 pages. 1993.

Vol. 664: M. Bezem, J. F. Groote (Eds.), Typed Lambda Calculi and Applications. Proceedings, 1993. VIII, 433 pages. 1993.

Vol. 665: P. Enjalbert, A. Finkel, K. W. Wagner (Eds.), STACS 93. Proceedings, 1993. XIV, 724 pages. 1993.

Vol. 666: J. W. de Bakker, W.-P. de Roever, G. Rozenberg (Eds.), Semantics: Foundations and Applications. Proceedings, 1992. VIII, 659 pages. 1993.

Vol. 667: P. B. Brazdil (Ed.), Machine Learning: ECML – 93. Proceedings, 1993. XII, 471 pages. 1993. (Subseries LNAI).

Vol. 668: M.-C. Gaudel, J.-P. Jouannaud (Eds.), TAPSOFT '93: Theory and Practice of Software Development. Proceedings, 1993. XII, 762 pages. 1993.

Vol. 669: R. S. Bird, C. C. Morgan, J. C. P. Woodcock (Eds.), Mathematics of Program Construction. Proceedings, 1992. VIII, 378 pages. 1993.

Vol. 670: J. C. P. Woodcock, P. G. Larsen (Eds.), FME '93: Industrial-Strength Formal Methods. Proceedings, 1993. XI, 689 pages. 1993.

Vol. 671: H. J. Ohlbach (Ed.), GWAI-92: Advances in Artificial Intelligence. Proceedings, 1992. XI, 397 pages. 1993. (Subseries LNAI).

Vol. 672: A. Barak, S. Guday, R. G. Wheeler, The MOSIX Distributed Operating System X, 221 pages. 1993.

Vol. 673: G. Cohen, T. Mora, O. Moreno (Eds.), Applied Algebra, Algebraic Algorithms and Error-Correcting Codes. Proceedings, 1993. X, 355 pages 1993.

Vol. 674: G. Rozenberg (Ed.), Advances in Petri Nets 1993. VII, 457 pages. 1993.

Vol. 675: A. Mulkers, Live Data Structures in Logic Programs. VIII, 220 pages. 1993.

Vol. 676: Th. H. Reiss, Recognizing Planar Objects Using Invariant Image Features. X, 180 pages. 1993.

Vol. 677: H. Abdulrab, J.-P. Pécuchet (Eds.), Word Equations and Related Topics, Proceedings, 1991. VII, 214 pages. 1993.

Vol. 678: F. Meyer auf der Heide, B. Monien, A. L. Rosenberg (Eds.), Parallel Architectures and Their Efficient Use. Proceedings, 1992. XII, 227 pages. 1993.

Vol. 679: C. Fermüller, A. Leitsch, T. Tammet, N. Zamov, Resolution Methods for the Decision Problem. VIII, 205 pages. 1993. (Subseries LNAI).

Vol. 682: B. Bouchon-Meunier, L. Valverde, R. R. Yager (Eds.), IPMU '92 – Advanced Methods in Artificial Intelligence. Proceedings, 1992. IX, 367 pages. 1993.

Vol. 683: G.J. Milne, L. Pierre (Eds.), Correct Hardware Design and Verification Methods. Proceedings, 1993. VIII, 270 Pages. 1993.

Vol. 684: A. Apostolico, M. Crochemore, Z. Galil, U. Manber (Eds.), Combinatorial Pattern Matching. Proceedings, 1993. VIII, 265 pages. 1993.

Vol. 685: C. Rolland, F. Bodart, C. Cauvet (Eds.), Advanced Information Systems Engineering. Proceedings, 1993. XI, 650 pages. 1993.

Vol. 686: J. Mira, J. Cabestany, A. Prieto (Eds.), New Trends in Neural Computation. Procedings, 1993. XVII, 746 pages. 1993.

Vol. 687: H. H. Barrett, A. F. Gmitro (Eds.), Information Processing in Medical Imaging. Proceedings, 1993. XVI, 567 pages. 1993.

Vol. 688: M. Gauthier (Ed.), Ada - Europe '93. Proceedings, 1993. VIII, 353 pages. 1993.

Vol. 689: J. Komorowski, Z. W. Ras (Eds.), Methodologies for Intelligent Systems. Proceedings, 1993. XI, 653 pages. 1993. (Subseries LNAI).

Vol. 690: C. Kirchner (Ed.), Rewriting Techniques and Applications. Proceedings, 1993. XI, 488 pages. 1993.

Vol. 691: M. Ajmone Marsan (Ed.), Application and Theory of Petri Nets 1993. Proceedings, 1993. IX, 591 pages. 1993.

Vol. 692: D. Abel, B.C. Ooi (Eds.), Advances in Spatial Databases. Proceedings, 1993. XIII, 529 pages. 1993.

Vol. 693: P. E. Lauer (Ed.), Functional Programming, Concurrency, Simulation and Automated Reasoning. Proceedings, 1991/1992. XI, 398 pages. 1993.

Vol. 694: A. Bode, M. Reeve, G. Wolf (Eds.), PARLE '93. Parallel Architectures and Languages Europe. Proceedings, 1993. XVII, 770 pages. 1993.

Vol. 695: E. P. Klement, W. Slany (Eds.), Fuzzy Logic in Artificial Intelligence. Proceedings, 1993. VIII, 192 pages. 1993. (Subseries LNAI).

Vol. 697: C. Courcoubetis (Ed.), Computer Aided Verification. Proceedings, 1993. IX, 504 pages. 1993.